Springer Textbooks in Earth Sciences,
Geography and Environment

The Springer Textbooks series publishes a broad portfolio of textbooks on Earth Sciences, Geography and Environmental Science. Springer textbooks provide comprehensive introductions as well as in-depth knowledge for advanced studies. A clear, reader-friendly layout and features such as end-of-chapter summaries, work examples, exercises, and glossaries help the reader to access the subject. Springer textbooks are essential for students, researchers and applied scientists.

More information about this series at http://www.springer.com/series/15201

William Bajjali

ArcGIS for Environmental and Water Issues

Springer

William Bajjali
Department of Natural Sciences
University of Wisconsin – Superior
Superior, WI, USA

ISSN 2510-1307 ISSN 2510-1315 (electronic)
Springer Textbooks in Earth Sciences, Geography and Environment
ISBN 978-3-319-87017-5 ISBN 978-3-319-61158-7 (eBook)
https://doi.org/10.1007/978-3-319-61158-7

Printed on acid-free paper

This Springer imprint is published by Springer Nature
The registered company is Springer International Publishing AG
The registered company address is: Gewerbestrasse 11, 6330 Cham, Switzerland

Preface

The study of geographic information systems (GIS) applications is enlightening, challenging, and very interesting. This workbook was created as a guide to students and professionals on the applications of GIS in the geoscience field. GIS applications is now considered an important course in the curriculum of undergraduate geoscience, environmental, and in some fields of engineering programs. It is a result of several years of experience in applying GIS technology to water resources and environment related problems. The databases and the applications used in the text reflect real world problems from different environmental settings that have been gathered from the author's works in the USA and the Middle-East. Each chapter presents a different set of scenarios and case studies that include an environmental problem that needs a solution. A step-by-step approach was adopted to provide answers and solutions to the problems for the scenarios presented.

The textbook is intended as an introductory course either at the undergraduate level or as a dual-level undergraduate/graduate course, and it can be used as a self-study workbook for professionals in the field of geoscience. It aims to teach students and professionals the various environmental disciplines of mapping and spatial analysis skills using ArcGIS and state of the art methodologies to acquire, visualize, and analyze data.

The book has 18 chapters that cover many topics related to GIS applications. Chapters 1 to 3 deal with introducing the concept of GIS, teach how to work with a variety of ArcGIS tools, classify data and perform different quantitative and qualitative classifications, and generate color-coded maps. Chapter 4 deals with data acquisition and data creation. It provides some reliable web pages where users can obtain different types of vector and raster data for carrying meaningful GIS projects. This chapter also discusses several ways to create one's own GIS data. It discusses how GIS features can be created by converting features from existing data or creating new data by drawing geographic features and adding attribute tables. Chapter 5 deals with coordinate systems, map projections, and re-projections. This chapter discusses the types of coordinate systems; geographic coordinate systems (GCS) (latitude-longitude) and various types of projected coordinate systems. The chapter explores the concept of the datum as it applies to the projection, projection on the fly, projection of the GCS into Universal Transverse Mercator (UTM), State Plane coordinate (SPC), various types of custom coordinate systems from the Middle East, re-projection, projecting the raster, and geo-referencing which is discussed through real-world examples. Chapter 6 describes vectorization work in the ArcScan tool. It is designed for vectorization, which means converting raster data to vector data. The process of vectorization (tracing) will be performed manually and automatically by using a rectified image from the Azraq Basin and converting it into a vector digital data file format. Chapter 7 deals with building a geodatabase, importing different data sources into the geodatabase, creating feature classes and feature datasets, and creating relationships between objects and a relationship class between a well feature class and two tables. Chapter 8 discusses data editing and topology. It discusses thoroughly the simple, advanced, and topological editing process using diverse real world examples. It demonstrates how to use simple editing tools to edit existing GIS features and how to fix some common digitizing errors such as overshoot and undershoots, generalize line feature, smooth polygon feature, merging, splitting, and reshaping feature classes. The chapter demonstrates data editing work using topology rules. These rules deal with understanding coincident and shared geometry, feature creation, building a geodatabase topology and identifying and fixing topology errors. Chapter 9 discusses geoprocessing, which refers to the tools and processes used to generate derived data sets from other data using a set of tools. Geoprocessing is a very important tool in ArcGIS software and plays a fundamental role in spatial analysis. Geoprocessing is discussed with examples based on extracting (clip, erase, and split) and combining features (merge, append, dissolve, and buffer), and combining geometries and attributes (union, intersect, and spatial join). Chapter 10 discusses site suitability and data modeling through two real world scenarios. Model 1 shows how to use different aspects of functionality in GIS to find the most suitable area for building a greenhouse at the Jordan University campus. Model 2 demonstrates how to find the best suitable location to build a nuclear power plant using the ModelBuilder in ArcGIS. Chapter 11 discusses geocoding in two scenarios; geocode based on the ZIP code case study from groundwater wells in Wisconsin, and geocoding the well addresses of owners by converting them into a point shapefile based on street address. Chapter 12 explains the use of raster analysis in GIS and is divided into three sections. Section 1 is about raster data download and raster dataset conversion, section 2 is about raster projection and processing raster datasets, and section 3 describes the terrain analysis. The user will perform different exercises of various GIS functions dealing with the raster analysis such as converting the neutral format (SDTS) into grid, creating hill-shade, contour, vertical profile, deriving view-shed, slope, aspect, mosaic images, clip image and much more. Chapter 13 deals with spatial interpolation techniques using groundwater salinity affected by salt intrusion. Trend analysis will be used to determine the direction of salinity distribution along the coast. The Global Polynomial Interpolation (GPI), Inverse Distance Weighting (IDW), and the Kriging interpolation techniques are used to study the effect of the dam on improving water quality along the coast. Chapter 14 explains the use of hydrology tools in the spatial analysis of watershed delineation using raster Digital Elevation Model (DEM). All the steps before watershed delineation such as using the flow direction, sink, fill, and flow accumulation tools are discussed. The catchment area of the Dhuleil-Halabat region has been delineated to estimate the amount of recharge. Chapter 15 discusses the use of geostatistical analysis to obtain meaningful information related to groundwater data in terms of distribution and patterns. The intention in the textbook is to focus on the GIS application rather than emphasizing the complex mathematics and statistics. Nevertheless, some of the tools such as measuring geographic distribution, analysis patterns, and mapping clusters will be implemented using groundwater resources. Chapter 16 discusses the proximity and network analysis techniques, which are important functions in GIS. They cover a wide range of topics that help answer many spatial questions related to the distance and movement along a linear route. The Near, Point Distance, and Desire line (Spider Diagram) tools are used to verify if the stored water behind the dam in the Dhuleil area infiltrates into the subsurface aquifer and improves its water quality. The Network Analyst is used to overcome natural barriers such as hills, lakes, or areas where there is absolutely no network of street system. The Network Analyst uses the actual distance that is associated with the street feature, which is an important feature in the application. This approach is more accurate than using the near function or spider diagram model. The Network Analyst will be used to find the amount of time required for a water truck to supply the towns in the Dhuleil region with portable water supplies in the summer time when there are water shortages. After finding the time required to supply the towns with water, users must find the actual path and time that the water truck will take to get from each individual well to each town. Chapter 17 explains the 3-D Analyst that is designed to do different types of analyses and make the map look real and easy to comprehend. The chapter discusses the display of 3-D in ArcCatalog, working with the 3-D Toolbar in ArcMap, and working with ArcScene. A scenario is discussed dealing with the movement of hydrocarbon leakage from an under storage tank of gasoline in 3-D. Chapter 18 explains how Mobile GIS (ArcPad) captures field data, where the users use a Trimble Juno T41 device. This chapter has a task that will be achieved in two parts. In part one, the user utilizes ArcGIS desktop to create a project containing the geodatabase and image needed for field collection. Part two instructs users to capture the locations of trees, sidewalks, and buildings on the main University of Wisconsin – Superior Campus.

Note on Data

The textbook includes three types of data that have been used in the exercises of the book's chapters. The first type of data presented is from actual field data gathered by the author and taken from his published work. The second type of data is manipulated based on real information gathered from different projects in the Middle-East. These data were modified with the aim of protecting the privacy and rules that govern these projects. The third type of data is public domain information and freely available from the Internet to any GIS user. Chapters that include public domain data in the exercises document the data sources.

Superior, WI, USA   William Bajjali

Acknowledgments

Many thanks to my students and colleagues from academia and my professional career, both past and present, who have worked with me over the years in the classroom conducting research and carrying out projects. Many of that work and the lessons learned were the motivation for authoring this text.

Contents

1 Introduction to GIS... 1

Introduction.. 1

What Is GIS?... 2

GIS Description 2

What Can a GIS Do?.. 2

Organization 3

GIS Infrastructure.. 3

GIS Principles... 3

Spatial Data Representation 3

Vector Data Model 4

Raster Data Model 6

Advantages and Disadvantages of the Raster and Vector Model 7

Raster Model... 8

Vector Model... 8

GIS Project 8

Lesson 1: Explore the Vector Data 8

Lesson 2: Explore the Raster Data 9

2 Working with ArcGIS: Classification 11

Introduction.. 11

Learning Objectives 12

ArcMap and ArcCatalog.. 12

Lesson 1: Working with ArcCatalog and ArcMap 12

Connect.. 12

Create a Thumbnail.. 13

Use the Search Window 14

Working with ArcMap 15

Change the Name of Layers 16

Symbolize the Layers 16

Find and Create a Bookmark 17

Create a Bookmark for the School North of the City of Chico 17

Using Relative Paths... 19

Save Map Document... 19

Lesson 2: Data Management.. 20

Change Names 20

Change the Symbols of the Layers and Show Map Tip 21

Label a Layer 22

Create a Bookmark.. 23

Create a Bookmark for the Groundwater Wellfield 23

Classification... 23

Classify the City Layer 23

3   Map Classification and Layout   25
Introduction...   25
Lesson 1: Creating Map and Data Classification   26
    Creating Geology Map Using Text Attribute   26
    Data Integration   26
    Creating Salinity Map Using Numeric Attribute   29

4   Data Acquisition and Getting Data into GIS   41
Section I: Introduction to the Acquisition Method   41
    Integrate Flat File Data into ArcGIS.................................   42
    Convert Degree, Minute, Second into Decimal Degree Using Excel........   42
    Integrate the Data Table into GIS   44
    Global Positioning System and GIS   44
    GPS Data Integration into ArcMap.....................................   44
    GPS Data Integration Using DNR Garmin Extension   45
    Download GPS Data and Integrate the Waypoints into ArcMap   45
    Add GPS Data to ArcMap Using ArcGIS Explorer Desktop   47
    Save the Waypoints Layer as a Layer Package (LPK)   48
    Integrate LPK into ArcMap   48
    Data Integration from the Internet   49
    First Download the Census Tract   50
    Second Download the Water Files   50
    Integrate Downloaded Data into ArcMap................................   51
    Aerial Photography and Satellite Images   51
    Download Image from the Internet.....................................   51
    Add Data from ArcGIS Online   53
Section II: Introduction to Feature Creating   55
    Creating a GIS Layer from Existing Feature   55
    Add New Features to the Existing Layer   57
    Digitizing on Screen   58
    Create New Polygon Shapefile   59
    Create the Polygon Shapefile   60
    Update the Attribute Table of the Building...........................   61
    Create New Line Shapefile   62
    Create New Point Shapefile   62
    Update the Digitized Map...   63
    Remove the Non-existing Buildings from the New Image.................   64
    Modify the E Building   65

5   Coordinate Systems and Projections.................................   67
Introduction...   67
Geographic Coordinate System (GCS)......................................   67
Map Projections   68
Projection and Distortion   68
Concept of the Datum..   70
Projection Parameters   71
Working with GCS   72
Projection on the Fly...   72
Define the Coordinate System and Datum   72
Apply the Projection on the Fly   73
Define the Projection...   74
How to Check That?..   75
Set the Environment   75
Projection the GCS into UTM Zone 15   76

Georeferencing   77
GIS Approach   78
   Set Environment   78
Raster Projection.   81
   Set the Environment   82
Project the Rectified Raster.   82
Datum Conflict   83
GIS Approach   83

6   **ArcScan**   89
Introduction.   89
Data Requirement   90
   Data Integration   90
   Make a Bi-level Image with 2-Classes by Reclassification   91
Use Reclassify Tool to Make the Image to Have Two Classes (0, 1)   91
Reclassification of the Image   92
Prepare the Image for Vectorization   93
Create a Blank Shapefile   94
Raster Cleanup   94
Remove the Undesired Objects in the Image   95
   First Approach.   95
   Second Approach   96
Vectorization.   97
   Interactive Vectorization (Vectorization Trace).   97
   Automatic Vectorization   100

7   **Geodatabase**.   103
Introduction.   103
Creating a Geodatabase.   103
   Create File Geodatabase   104
   Create a Feature Class and Assign for it WGS84-GCS   104
   Capture the Feature Classes Using an Image   105
   Create Feature Dataset and Import Shapefiles into It   107
   Import the Stream, StudyArea, and Dam into the Water Feature Dataset   107
Create a Relationship Class   108
   Create Personal Geodatabase   109
   Integrate Shapefiles and Tables into Personal Geodatabase   109
   Create Relationship Class.   110
   Create Relationship Class in Jizzi.gdb Between Table1 and Table2.   112
   Find the Wells that have Salinity Higher than 500 mg   114
   Projection and Datum Conflict   115
   GIS Approach   116

8   **Data Editing and Topology**.   117
Introduction.   117
I: Simple Editing.   118
   Delete Function.   119
   Move Function   119
   Split Function   120
   Reshape Function   121
   Modify Feature   122
   Update the Area and Perimeter Field in the Geology Attribute Table   124
   Merge Function.   125

II: Advance Editing.. 126
    Fixing Overshoots and Undershoots................................. 126
    Correct Overshoots of the Street.................................. 128
    Generalize a Stream Feature....................................... 128
    Smooth a Lake Feature... 129
III: Topological Editing Using Geodatabase............................. 130
    Fix Fault System Using Topology................................... 130
    Editing Using Topology 131
    Create Geodatabase.. 131
    Building Topology and Set the Rules 132
    Fixing Dangles in ArcMap Using First Topology Approach 133
    Topology Tool... 133
    Fixing Dangles in ArcMap Using Second Topology Approach 135
    Fix Watershed Using Topology...................................... 136
    Build Topology Rule to Make Two Watershed Layers Cover Each Other 137

9   Geoprocessing 141
    Introduction.. 141
    GIS Approach 142
        Dissolve.. 142
        Clip Tool 144
        Intersect Tool.. 145
        Merge Tool 146
        Buffer Tool and Select by Location 147
        Convert the Graphic into Feature 149
        Erase Tool 150

10 Site Suitability and Modeling...................................... 153
    Introduction.. 153
    Model 1 153
        The Criteria to Build the Greenhouse 154
        GIS Approach 154
        Overlay Analysis: Union....................................... 156
        Select Tool... 156
        Area Calculation 157
        Geoprocessing Model for Spatial Analysis...................... 158
    Model 2 159
        Find Best Suitable Location to Build Nuclear Power Plant 159
        The Criteria to Find Suitable Location to Build Nuclear Power Plant 159
        Building the Geoprocessing ModelBuilder 159
        Create a New Toolbox.. 160
        Select Tool... 164
        Union Tool 165
        Erase Tool 167
        Validate and Run the ModelBuilder in Model Window 168
        Add Model Name 170
        Set Model Parameters and Run ModelBuilder in Model Tool 171
        Run the Model Tool 174

11 Geocoding 177
    Introduction.. 177
    Geocoding Based on Zip Code....................................... 177

GIS Approach 178
Create Address Locator. 179
Geocode the Addresses 180
Create Table with the Nitrate Information 181
Symbolizing 182
Geocoding Based on Street Address. 183
Integrate Excel Table. 183
Create Address Locator. 184
Test Your Address Locator 185
Examine the Geocoding Results. 187
Match the Unmatched Addresses 187
Rematch the Address. 187
First Address (2066 Fisher AV). 188
Second Address (666 20TH AV). 188

12 Working with Raster. 191
Introduction: Raster Format 191
Feature Representation in Raster Format 191
Section 1: Data Download and Display in ArcMap 192
Section 2: Projection and Processing Raster Dataset 192
Section 3: Terrain Analysis 192
Section 1: Data Download and Display in ArcMap 192
Download DEM Image from USGS Webpage 192
Explore the DEM Image. 194
Convert Image from Float to Integer 196
Int Tool 196
Section 2: Projection and Processing Raster Dataset 197
Project the DEM of an Area in the Amman-Zarqa Basin 197
Clip an Image 198
Merge Raster Datasets (Mosaic). 199
Resample an Image. 200
Classify an Image 201
Convert Vector Feature into Raster. 201
Section 3: Terrain Analysis 203
Create Hillshade 203
Create Contour for Dhuleil DEM 204
Create Vertical Profile. 205
Create Visibility Map 206
Add a Height to an Observation Point 207
Create Line of Sight 208
Slope and Aspect. 209
Classify the Slope into Six Classes. 210
Reclassify Slope and Aspect. 211
Combine Two Images: Slope and Geology. 214
Calculate the Percent of the Area That Suitable for Installation the Lysimeter . . . 217

13 Spatial Interpolation. 219
Introduction. 219
Method of Interpolation 220
Trend Surface Analysis. 220
Inverse Distance Weighting (IDW). 220

Global Polynomial (GP) 221
Kriging 221
Data and Coordinate System.. 222
Density of Groundwater Well . .'..................................... 222
GIS Approach 223
Symbolize the Wells Based on TDS 223
Trend Analysis 225
I: Global Polynomial Interpolation 226
Convert the GPI Predicted Map into ESRI Grid and Clip
It Using the Mask Technique 228
Classify the GPI Map 228
II: Inverse Distance Weighting (IDW) 229
Convert the IDW Predicted Map into ESRI Grid and Clip
It Using the Mask Technique 230
Classify the IDW Map 231
Interpolation Using Kriging 232
Convert the Kriging Predicted Map into ESRI Grid and Clip
It Using the Mask Technique 234

14 **Watershed Delineation** 235
Introduction... 235
Flow Direction 236
Flow Accumulation... 236
Stream Link... 236
Delineate the Watershed 237
Step 1: Run the Flow Direction Tool 237
Step 2: Identify the Locations of the SINK (Sink Tool) 238
Step 3: Run the Fill Tool.. 239
Step 4: Run the Flow Direction tool 239
Step 5: Create a Flow Accumulation Raster 240
Step 6: Create Source Raster to Delineate Watershed 241
Source Raster 241
Step 7: Delineate Watershed....................................... 242
Point-Based Watershed.. 242
Convert Pourshed Raster to Vector 243
Calculate the Recharge amount of the Dhuleil_Watershed.............. 244

15 **Geostatistical Analysis** 247
Introduction... 247
Measuring Geographic Distribution Toolset............................ 247
Calculate Mean Center with and Without Weight..................... 248
Standard Distance and Mean Center................................. 251
Analyzing Pattern Toolset... 253
Identifying Pattern Based on Location 253
Identify Pattern Based on Values (Getis-Ord General G).............. 257
Spatial Autocorrelation (Moran's I) 259
Mapping Clusters 262
Cluster and Outlier Analysis (Anselin Local Moran I)................ 262
Hot Spot Analysis (Getis-Ord GI*)................................. 264

16 **Proximity and Network Analysis**.................................... 267
Introduction... 267
Proximity Analysis in Vector Format 267

GIS Approach to Solve Scenario 1 268
   Buffer the WWTP in the Samra Region 269
   Select By Location 270
   SQL Statement 270
   Buffer the Stream in the Region 272
   Definition Query 274
   Near Tool 274
   Desire Lines (Spider Diagram) Tool 275
GIS Approach to Solve Scenario 2 276
   Use the Multi-Ring Buffer Around the Hay Arnous Town 276
GIS Approach to Solve Scenario 3 278
   Proximity Analysis in Raster Format 278
   Point Distance Method 278
   GIS Approach to Solve Scenario 3 279
   Point Distance Tool 280
   Definition Query 282
   Well Classification 282
Network Analyst 284
   GIS Approach 284
   Create Network Dataset 287
   Build the Network 288
   New Service Area 289
   Add Locations Tool 290
   Calculate the Service Area with Certain Travel Time 290
   Calculate the True Path and Total Time between the Wells and Each Town 292
   GIS Approach 293
   Run Add Locations Tool Between Facilities and Well_Supply 294
   Run Add Location Tool Again 295
   Show Route Direction 296

17   3-D Analysis 297
  Introduction 297
  Z Values 297
  Raster 297
  Triangulated Irregular Network (TIN) 298
  3-D Features 298
  Lesson 1: Working with 3-D in ArcCatalog 298
   Preview Raster (DEM) and Vector in ArcCatalog 298
   Does the Shapefile Contain 3D Features? 300
   Create a Layer File for the DEM of Duluth in ArcCatalog 300
   Display Duluth.lyr in 3D Using the Base Height of Duluth DEM 300
   Change the Color of the Duluth Layer File 301
   Add Shading to the Duluth Layer File 301
   Create a 3D Thumbnail to the Duluth Layer File 302
  Lesson 2: Working with 3-D Toolbar in ArcMap 303
   Create Single Contour 304
   Find the Elevation of the Rain Stations Using the "Interpolate Point" 304
   Create a Profile Graph for the Stream Using the "Interpolate Line" 305
  Lesson 3: Working with ArcScene 306
   Create TIN from Contour Line 306
   Change the Symbols of the TIN 307
   Drape and Extrude Layers onto Dhuleil_TIN 308

Applying the 3D Symbol to the Tree Layer 309
Vertical Exaggeration 310
Navigate and Fly... 310
Create an Animation and Video 310
Time Tracking.. 312

18 Mobile GIS Using ArcPad 315
Introduction.. 315
Part 1: Prepare the Data in the Office for ArcPad to Perform Field Work 315
Prepare Raster Data for ArcPad Desktop 316
Create File Geodatabase in ArcMap.................................. 318
Import the Image (Holden.tif) into the UWS File Geodatabase.............. 320
Using ArcPad Data Manager Tool to Transfer Data from ArcGIS
into the Mobile GIS 321
Starting ArcPad in Trimble Juno T41 325
GPS Setting in Trimble... 328
Activate the GPS... 329
Part 2: Capturing Data Using Trimble and ArcPad in the Field 329
I: Capturing Data Using the Designed File Geodatabase 329
Capturing the Building, Sidewalk, and Trees in the Designated Study Area 330
Get Data from ArcPad to ArcGIS 332
Editing the Data of the Holden Building 333
II: QuickProject.. 334
Activate the GPS in Trimble.. 335
Capturing GPS Data (Point, Line, Polygon).......................... 335
View the Picture That Was Captured in ArcPad in the Field Using Trimble 339
Create Map Hyperlink.. 341

Appendix A: Data Source Credits 343

Appendix B: Task-Index.. 347

Introduction

GIS (Geographic Information System) has great value in our time as it is a comprehensive information system evolved and still developing parallel with the advancing technology. This era of the human kind is characterized as an information age, where the whole world is experiencing and interacting with a new revolution that change our traditional way to look at the things and do a business in completely different approach. The emphasis is on the technology and its use in every activity that range from agriculture, industry, business, social, research and education. The advancement in technology changed our world and our approaches to meet our need to rely completely on the technology and data. The value of information in our time becomes vital and important for development. GIS itself is an important module of the information system. The economy of all industrial countries and many other nations all over the globe, become more dependent on services. This means that the current economy rely more and more on computers, networking, accurate information and data. This shift required a mass of skilled labors that are capable to deal with the technology and data processing.

GIS technology is not an exception when it comes to its use in water resources, geology, and environmental related problems. It is a powerful tool for developing solutions for many applications ranging from creating a color coded geological map, interpolating the water quality of groundwater aquifer to managing water resources on a local or regional scale.

Water is the most precious and valuable resource and vital for socioeconomic growth and sustainability of the environment. In some arid countries, the water resources are limited, scarce, and mainly sourced from groundwater. In some Middle Eastern countries surface water is limited to few river systems and intermittent streams that are associated with rain during the winter time. Precipitation is vital and the primary source of recharge for various groundwater aquifers in these regions.

Groundwater in the region has been utilized through wells tapping various water-bearing formations to provide more fresh water to supply the increased demand water supply and irrigation. This practice negatively affected the whole hydrogeological setting of the basins. For example, total water withdrawal in the region (Israel, Jordan, and Palestinian territories) in 1994 was about 3050 million cubic meters. The estimated total renewable water supply that is practically available in the region is about 2400 million cubic meters per year. The water deficit is being pumped from the aquifers without being replenished. This practice caused the groundwater level to decline dramatically in some well fields, up to 20 m, which caused some major springs in 1990 to cease completely in Azraq basin, Jordan.

Therefore, management of water resources has become a major effort for governments in the region. Various ministries, water institution, and private companies worldwide are using the GIS as a tool to manage water resources in their countries. The GIS can be used to capture data and developing hydrologic dataset for all components of water resources. Understand the region's hydrology, map sources of contaminations, prepare water quality and water-rock interactions maps, delineate the watershed areas, and much more.

Electronic Supplementary Material: The online version of this chapter (https://doi.org/10.1007/978-3-319-61158-7_1) contains supplementary material, which is available to authorized users.

What Is GIS?

GIS is an information technology system that stores all the digital data in one location, retrieving and analyzing the data quickly and efficiently in a color coded map format. The map creation depends on database and information. Users can use various types of databases to create different layers in one display. This will add an advantage to the data analysis by providing the spatial relationship between layers and will reveal any hidden relationship that the user can use for facilitating the analysis and strengthen the finding. For example, soil layers can be viewed with elevation layers and water table depth of an aquifer, in order to select the best location for building a landfill.

In previous years, all maps were created manually. A cartographer would use simple tools and paper to draw a map. This work was tedious and time consuming and if an error took place, the cartographers had to repeat all of their work and start over. This map was static in a way that any new development couldn't be added.

With the advancement in technology, a GIS map is more dynamic, can be modified in very little time, and can be stored, displayed, and printed out quickly and efficiently. GIS is a new methodology in science and applications, it is a new profession and a new business.

GIS refers to three integrated parts.

1. Geographic: The geographical location of the real world (coordinate system)
2. Information: The database
3. Systems: The hardware and software

GIS Description

A GIS is a computer-based tool that helps us visualize information with patterns and relationships that aren't otherwise apparent. The ability to ask complex questions about data and analyze many features at once, then instantly see the results on a map is what makes GIS a powerful tool for creating information. GIS can be used in many disciplines such as resource management, criminology, urban planning, marketing, transportation, etc. Primarily GIS is used for scientific analysis but is now being implemented in other disciplines.

What Can a GIS Do?

A GIS performs six fundamental operations that make it a useful tool for finding solutions to real-world problems. Throughout this course, you will gain experience with the ArcGIS tools used for these operations.

1. **Capture data:** You can add data from many sources to a GIS, and you can also create your own data from scratch. You will learn about getting data into a GIS in Chap. 4.
2. **Store data:** You can store and manage information about the real world in ways that makes sense for your application. You will learn about organizing data in Chap. 3, 4, and 7.
3. **Query data:** You can ask complex questions about features based on their attributes or their location and get quick results. You will gain experience with querying in Chap. 7, 8, 12, and 16.
4. **Analyze data:** You can integrate multiple datasets to find features that meet specific criteria and create information useful for problem solving. You will perform analysis in Chap. 2, 4, 5, 6, 8, 9, 11, 12, 13, 14, 15, 16, 17 and 18.
5. **Display data:** You can display features based on their attributes, a powerful feature you'll come to appreciate. You will learn how to symbolize features in different ways in Chap. 2 and 3.
6. **Present data:** You can create and distribute high-quality maps, graphs, and reports to present your analysis results in a compelling way to your audience. You will learn how to design an effective map in Chap. 3.

GIS is a computerized system that deals with spatial data in terms of the following:

1. *Storage*: Digital and database storage.
2. *Management of Data*: Integration of the database into the GIS system.

3. *Retrieval*: The capacity to view the various database data formats.
4. *Conversion*: Convert different sets of data from one form to another.
5. *Analysis*: Manipulating data to produce new information.
6. *Modeling*: Simplifying the data and its process.
7. *Display:* Presenting the output works.

Organization

GIS is a complete system that consists of sophisticated hardware and software. It performs many integrated functions:

1. GIS accepts data from multiple sources, which can be in a variety of formats. For example, if you are dealing with ArcGIS you can work not only with Shapefile, but also with Coverage (ARC/INFO format), Geodatabase, DXF (AutoCAD format), DBS (database system), and other types of database and digital formats.
2. Data types include the following:
   - Maps (Tiff, Jpeg, etc.)
   - Images from aircraft and satellite
   - Global Positioning System (GPS): (Coordinates, elevation)
   - Text data (report and text)
   - Tabular data (excel file)

GIS Infrastructure

1. Hardware: The machine where the GIS can be run (computer, digitizer, plotter, printer).
2. Software: The program needed to run the GIS (ArcGIS and its extensions)
3. Data: The digital and database (information)
4. Organization and People: This is the most important part of the GIS structure. The GIS is too important and so costly that it cannot be considered just equipment. It requires organization and staff to utilize this technology. Unfortunately many organizations treat the GIS as equipment rather than an important analysis tool.

GIS Principles

1. The computer is an unavoidable technology in our time. We are living in the digital age, which has become an important element in nearly all professions.
2. Computer training in most scientific disciplines is essential. Without this technology all professionals will be handicapped.
3. The GIS is an inevitable technology that will be used in all scientific fields. The GIS has become the accepted and standard means of using spatial data.
4. GIS is more *Accurate Flexible*, *Object Efficient*, and *Rapid Fun* comparing with the traditional method of spatial data inventory.
5. GIS is replacing traditional cartography. Much of traditional "pen and ink" cartography done by skilled draftsperson and artist is being replaced by GIS.
6. GIS is opening new horizons. New mode of analysis and applications are constantly discovered.

Spatial Data Representation

Spatial data is a fundamental component in any GIS environment. The data is based on the perception of the world as being occupied by features. Each feature is an entity which can be described by its attribute or property, and its location on earth can be mapped using a spatial reference. The most common representation of spatial data that measures the landscape is using discrete data (vector model) and continuous data (raster model). The data models are a set of rules used to describe and represent real world features in a GIS software.

Vector Data Model

The vector data model is a representation of the world of distinct features that have definite boundaries, identities, and has a specific shape using point, lines, and polygons. Vector data is structured with two specific elements (node, vertex) and coordinates. This model is useful for storing data that has discrete boundaries, such as groundwater wells, streams, and lakes. Each entity has a dimension, boundary and location. For example, a well has a specific measurement and its location can be described using a coordinate system such as latitude-longitude. The following represents the three fundamental vector types that exist in GIS.

Point: A point entity is simply a location that can be described using the coordinate system (longitude, latitude or X, Y). The point has no actual spatial dimension and has no actual length and width but has a specific location in space (single coordinate pair). Point can be represented by different symbols. Points generally specify features that are too small to show properly at a given scale. For example, buildings, schools, or a small farm at a scale of 1:25,000 can be represented as a point.

Figure 1.1 shows the location of five groundwater wells with each well representing a point feature. Table 1.1 shows the coordinate locations of the wells in (X, Y). The coordinate system allows users to integrate the wells into GIS and make them subject to mapping. The well feature is associated with an attribute table. The attribute of each well has information related to the depth and the yield of each well.

Line: A spatial feature that is given a precise location that can be described by a series of coordinate pairs. Each line is stored by the sequence of the first and last point together with the associated table attribute of this line. Line is one dimensional feature and has length but no width. Lines are a linear feature such as rivers, pipelines, and fences. The more points used to create the line, the greater the detail. The recent requirement that the line features include topology, which means that the system stores one end of the line as the starting point and the other as the end point, giving the line "direction".

Fig. 1.1 Point feature representation

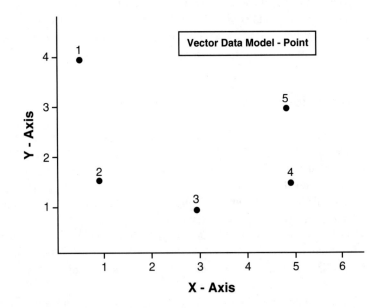

Table 1.1 Attribute table of point feature

No.	Well	X	Y	Well	Depth (m)	Yield (m³/h)
1	WAJ-1	0.5	4.0	WAJ-1	78	90
2	WAJ-2	1.0	1.5	WAJ-2	48	68
3	WAJ-3	2.95	0.9	WAJ-3	35	54
4	WAJ-4	4.95	1.5	WAJ-4	58	75
5	WAJ-5	4.90	3.0	WAJ-5	55	75

Fig. 1.2 Line feature
representation

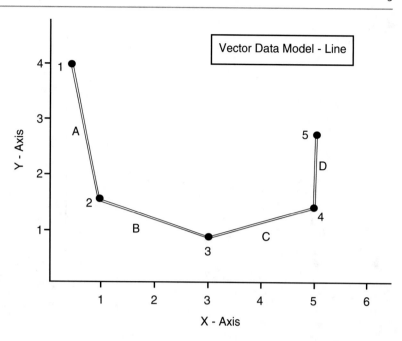

Table 1.2 Attribute table of line feature

Node No	X	Y
1	0.5	4.0
2	1.0	1.5
3	2.95	0.9
4	5.0	1.5
5	5.1	2.8
Line	1st Node	Last Node
A	1	2
B	2	3
C	3	4
D	4	5
Line	Length	Discharge
A	25.5	5
B	20.6	4
C	20.6	4
D	15.0	3

Figure 1.2 shows four pipelines (A, B, C, and D). Each pipeline is represented by a line that has its first and last node to distinguish its location. Each line has attributes of length and discharge. Notice that each node has coordinates (X, Y) stored in another table (Table 1.2).

Polygon: The polygon is an area fully encompassed by a series of connected lines. The first point in the polygon is equal to the last point. Polygon is a 2-D feature with at least three sides and because lines have direction, the area that falls within the lines compromise the polygon and the perimeter can be calculated. All of the data points that form the perimeter of the polygon must connect to form an unbroken line. Polygons are often an irregular shape such as parcels, lakes, and political boundaries.

Figure 1.3 shows polygon A, which represents an agricultural field. The polygon has its first and last node in node number 1 to settle its location. Node 1, vertexes 2, 3, 4, 5, and 6 have coordinate (X, Y) that is stored in another table (Table 1.3). Aside from location attributes, the polygon has associated attributes of area and crop.

Features on maps have spatial relationships which shows how those features are related to each other in space. The most important spatial relationships are:

Fig. 1.3 Polygon feature representation

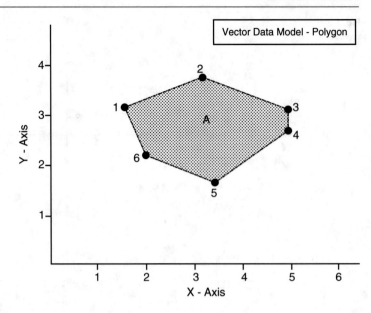

Table 1.3 Attribute table of polygon feature

Node No 1 and vertexes	X	Y
1	1.6	3.1
2	3.2	3.8
3	5.0	3.1
4	5.0	2.6
5	3.4	1.8
6	2.0	2.2
Polygon	Node	
A	1, 2, 3, 4, 5, 6	
Polygon	Area	Crop
A	520	Tomato

1. **Distance**: This measures the distance from one feature to another in the GIS map. The distance concept is an important relationship as the distance between features can be measured in any unit regardless of the map's coordinate system.
2. **Distribution**: This is the collective location of features where relationships can show the feature among themselves or their spatial relationships with other features in the map.
3. **Density**: This is the number of features per unit area or simply how close features are to each other.
4. **Pattern**: This is the consistent arrangement of a feature.

Raster Data Model

A representation of an area or region as a surface divided into a grid of cells (Fig. 1.4). It is useful for storing data that varies continuously, as in an aerial photograph, a satellite image, a surface of humidity, or digital elevation model (DEM) (Fig. 1.5).

Figure 1.4 depicts the basics of the raster data model. The cell is the minimum mapping unit and the smallest size at which any landscape feature can be represented.

These cells in the raster dataset are used as building blocks for creating points, lines, and polygons. In the raster data model, points, lines, polygons are represented by grid cells. The location of each cell in the grid is determined by two things: (1) the origin of the grid (the upper left-corner, which is (0, 0) and the resolution (size of the cell). The resolution is determined by measuring one side of the square cell. For example, a raster model with cell representing 5 m by 5 m (25 m^2) in the real

Fig. 1.4 Characteristics of raster data structure

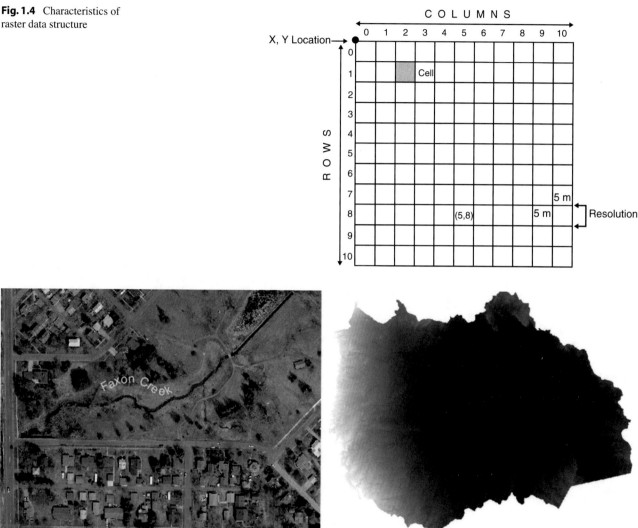

Fig. 1.5 Aerial photograph of Faxon Creek, Superior, WI (*left*) and DEM Jafr, Jordan (*right*)

world would be said to have a spatial resolution of 5 m. Each cell in the raster carries a single value, which represents the characteristic of the spatial phenomenon at a location denoted by its row and column. The precision of raster data is ruled by the resolution of the grid data set. The data type for that cell value can be either integer or floating-point.

The raster model will average all values within a given cell to yield a single value. Various techniques are used to assign cell code such as presence-absence, cell center, dominant area, and percent coverage. The more area covered per cell, the less accurate the associated data values. The area covered by each cell determines the spatial resolution of the raster model from which it is derived. Raster coding produces spatial inaccuracies as the shape of features is forced into an artificial grid cell format. Therefore, there is no way to know where any small feature occurs within the cell as the location according to the raster format is simply the entire cell. If the raster cell representing 100 m by 100 m, and the cell represents a well that has 0.5 m diameter. The cell in this case represents the well and this makes the raster format imprecise.

Advantages and Disadvantages of the Raster and Vector Model

There are several advantages and disadvantages for using either the raster or vector data model for storing and displaying spatial data.

Raster Model

Advantages
1. Simple data structure
2. Efficient for remotely sensed or scanned data
3. Simple spatial analysis procedures

Disadvantages
1. Requires greater storage space on a computer
2. Depending on pixel size, graphical output may be less pleasing
3. Projection transformations are more difficult
4. Difficult to represent topological relationships

Vector Model

Advantages
1. Data can be represented in its original resolution without generalization
2. Requires less disk storage space
3. Topological relationships are readily maintained
4. Graphical output closely resembles hand-drawn maps

Disadvantages
1. More complex data structure
2. Inefficient for remotely sensed data
3. Some spatial analysis procedures are complex and process intensive
4. Overlaying multiple vector maps is often time consuming

GIS Project

To carry a GIS project, users need to integrate spatial data into the GIS software where the data can have a vector or raster dataset. GIS data comes from many resources

1. Hard copy maps
2. Digital files
3. Imagery
4. GPS
5. Excel, text delimited, and dbf files
6. Reports

The GIS analysis is based on the database, which is powerful and important in GIS.

Lesson 1: Explore the Vector Data

Layers in vector data format in GIS can be a point, line and polygon, either in shapefiles, coverage, or geodatabase feature classes. These layers can be integrated in ArcMap and classified and symbolized by different symbols. In this lesson you are going to display Newton Creek that starts from Murphy Oil Inc. (Superior, WI) and discharges into Hog Island including the 5-sampling sites.

1. Launch ArcMap[1]

2. Click Add Data  in the Standard Toolbar

3. In the Add Data Dialog box click the Connect To Folder  button

4. Browse to Ch01/click OK
5. D-click Data folder/open Shapefile folder and highlight **HugIsland.shp**, **MurphyOil.shp**, **NewtonCreek.shp**, and **SamplingSite.shp**
6. Click Add

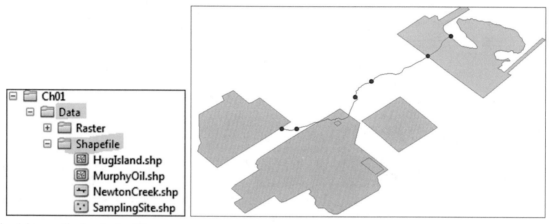

Result: The 4-layers display in ArcMap

Lesson 2: Explore the Raster Data

A raster map is full of continuous data like a photograph and uses different formats. In this lesson you are going to use a Digital Elevation Model (DEM) of Jafr basin (Jordan). It is a depression basin located in the east of the country and has groundwater resources used for irrigation.

7. Insert menu/click Data Frame

8. Click Add Data  in the Standard Toolbar
9. Navigate to \\Ch01\Data\Raster and integrate **jafr** (DEM)

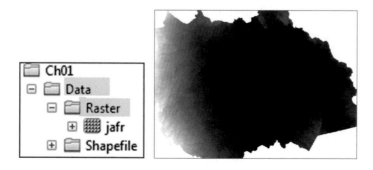

---

[1] From the window taskbar/Start/All Programs/ArcGIS/ArcMap.

10. In the Table of Content (TOC)/R-click Jafr image/Properties/click Source tab

Note: the Raster property includes four sections that provide detailed information about the Jafr DEM

- **Raster Information**: This shows that the DEM has 15,689 columns and 11,633 rows. It shows that the image consists of 1 band and the cell size (resolution) is 10 × 10 m. It also indicates that the pixel type is a signed integer, meaning that the raster has an attribute table and can be opened in ArcGIS. The Pixel depth of the raster is 16 Bit.
- **Extent**: this shows the coordinate extent in meter.
- **Spatial Reference**: shows that the coordinate of the Jafr DEM is registered in Palestine_1923_Palestine_Belt (customized UTM projection) and the datum is D_Palestine_1923. It also provides information about the parameters of the projection such as the false easting and false northing.
- **Statistics**: this shows the minimum and maximum elevation of the area.

Raster Information	
Columns and Rows	15689, 11633
Number of Bands	1
Cell Size (X, Y)	10, 10
Uncompressed Size	696.22 MB
Format	GRID
Source Type	Generic
Pixel Type	signed integer
Pixel Depth	16 Bit

Extent	
Top	1013205.07347
Left	191250.731268
Right	348140.731268
Bottom	896875.073475

XY Coordinate System	Palestine_1923_Palestine_Belt
Linear Unit	Meter (1.000000)
Angular Unit	Degree (0.0174532925199433)
False_Easting	170251.555
False_Northing	1126867.909
Central_Meridian	35.21208055555556
Scale_Factor	1
Latitude_Of_Origin	31.73409694444445
Datum	D_Palestine_1923

Abbreviations

D-click   Double Click
R-click   Right Click
TOC   Table of Content

Introduction

The benefit from the use of geographic information system (GIS) software is tremendous and range from managing the transportation in a dense city, to finding and modeling sophisticated environmental problems. GIS is used all over the world to achieve various tasks from managing the environment, and offering better service. It allows the user to carry research, and studies practically everything such as land, climate, environment, natural resources, population, etc.

Chapter 2 introduces the fundamental concepts of GIS, and the major functionality contained in ArcGIS Desktop. You will work with a variety of ArcGIS tools, and you will learn how to create color coded maps, query, and solve a variety of spatial problems.

ArcGIS desktop is a collection of software products for building comprehensive GIS. ArcGIS is the collective name for three products: ArcView, ArcEditor, and ArcInfo. These products have the same interface and share some of their functionality.

GIS is made up of layers or themes that make maps in GIS. You can add as many layers as you want, and the layers may contain features or images.

This section will focus on three components of the most advance ArcGIS desktop: ArcMap, ArcCatalog, and ArcToolbox.

ArcMap make maps from the layers of spatial data, choose colors, symbols, query, analyze spatial relationship, and design map layout.

ArcCatalog browse spatial data contained on the hard disk, network, or internet, search for spatial data, preview data, create features and metadata.

ArcToolbox use many tools to perform different types of analysis ranging from converting spatial data from one format to another, projection data, interpolation, analysis, and many others (you will learn about ArcToolbox in the next chapters).

The lessons in this chapter provide an overview of the basic GIS concepts and standard ArcGIS functions. You'll work with the ArcGIS software and will familiarize yourself with its main components, ArcMap, and ArcCatalog.

Electronic Supplementary Material: The online version of this chapter (https://doi.org/10.1007/978-3-319-61158-7_2) contains supplementary material, which is available to authorized users.

Learning Objectives

A student who completes this module will be able to:

1. Explore a GIS map and get information about map features
2. Preview geographic data and metadata
3. Add data to a map
4. Describe the structure of a GIS map
5. Explain how a GIS represents real-world objects
6. Change the way features are drawn on a map
7. Access feature information in different ways
8. Describe spatial relationships of map features
9. Describe how GIS can be used to solve problems

ArcMap and ArcCatalog

Using the ArcMap allows the user to control the whole map, and do changing, modification, and updating, at any time, within a very short period of time. ArcMap makes the map be dynamic, in a way that it is subject to any change, in term of color, symbols, classification, and layout. You can perform and distance measurement, in any unit you desire; regardless the map unit of the map.

You can zoom in and out to see different areas with more or less detail, you can decide what features you want to see and how they are symbolized. But most important is that you can retrieve the database, of the features, displayed on the map.

In this topic, you'll learn some of the concepts to which GIS maps are based. But first, you'll do an exercise to see just how easy they are to use and explore.

The ArcCatalog application organizes and manages all GIS information, such as maps, globes, data sets, models, metadata, and services. It is an application for managing your data in term of copy, delete, review, browse, search, and other functionality.

It includes tools to

1. Browse and find geographic information
2. Record, view, and manage metadata
3. Define, export, and import geodatabase schemas and designs
4. Search and browse GIS data, on local networks and the Web

Lesson 1: Working with ArcCatalog and ArcMap

Scenario 1: You are new GIS employee at the city of Chico, California, and you have been asked by your superior to do the following:

1. Launch ArcCatalog
2. **Connect** to your directory in ArcCatalog \\Ch02\
3. Create a **thumbnail** for all the layers
4. Use the **search** button on California
5. **Integrate** all the layers into ArcMap
6. **Change** the name of the layers, and give them a proper name
7. **Symbolize** the school
8. Create a **bookmark** for the school in the north of city of Chico
9. Shift your work between **Data View** and **Layout View**

Connect

1. Start ArcCatalog
   Note 1: In Window/Start/All Programs/ArcGIS/ArcCatalog 10.5 **OR**
   Note 2: R-click ArcCatalog/Send to/Desktop (to create shortcut permanently in desktop

2. Connect to your directory in ArcCatalog

    (a) Click "Connect To Folder" button

    (b) Browse to the following directory (\\Ch02)

    (c) Highlight Ch02 and click "OK"

Create a Thumbnail

Thumbnail allows you to use pictures (images) instead of shapes for variable and tool elements in a model diagram. Thumbnails draw quickly because they are snapshots; the data isn't displayed when you see a thumbnail.

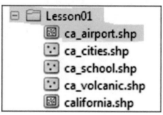

1. In the Catalog Tree, click the plus sign to open Q1

2. Highlight the file "**ca_airport.shp**"

3. Click "preview tab"

4. Click "Create Thumbnail" button   in the Geography Toolbar

5. Repeat the previous steps to create thumbnails to the rest of the files

6. Highlight the folder (\\Q1)

7. Click "Contents" tab to see all the thumbnails

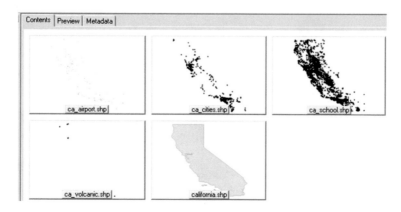

Note: make sure you are clicking the Thumbnail button in the Standard toolbar

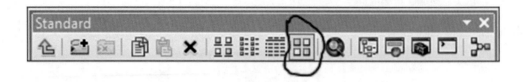

Use the Search Window

1. Click the "**Search**" window icon in the Standard toolbar

2. The Search dialog box display
3. Click on the "**Search Options**" icon on the ArcGIS **Search** Window

5. The dialog box of "**Search Options**" display
6. Click Add
7. Browse to the directory \\Ch02\Data\Q1
8. Click the "**Index New Items**" tab
9. Click Apply/OK when finishing indexing
10. Write in the Local Search the word "**Ca**" and click search
11. Four files displayed ca_airport, ca_cities, ca_school, and ca_volcanic

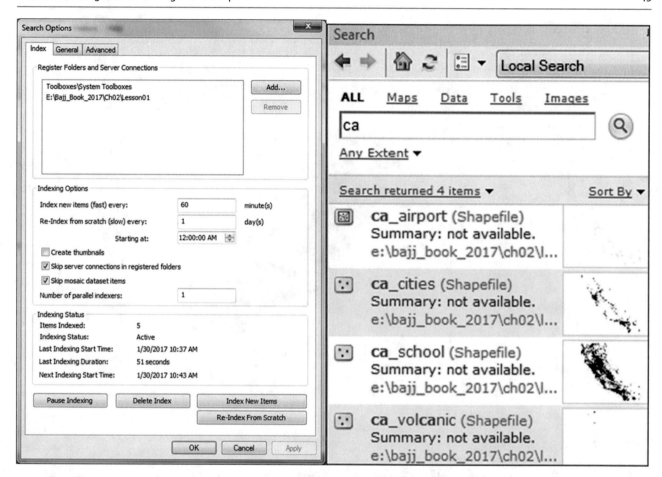

12. Click in the link below the Ca_airport file this will highlight the Ca_airport in the table of content of Catalog Tree in the left panel
13. Close the search window

Working with ArcMap

1. Start **ArcMap** through the "launch ArcMap" button in Standard toolbar in **ArcCatalog**

   OR

   a. From the window taskbar/Start/All Programs/ArcGIS/ArcMap

2. Minimize the ArcCatalog and ArcMap so you can see both of them
3. Drag the 5-layers one by one (ca_airport.shp, ca_cities.shp, ca_school.shp, ca_volcanic.shp and california.shp) from ArcCatalog to ArcMap

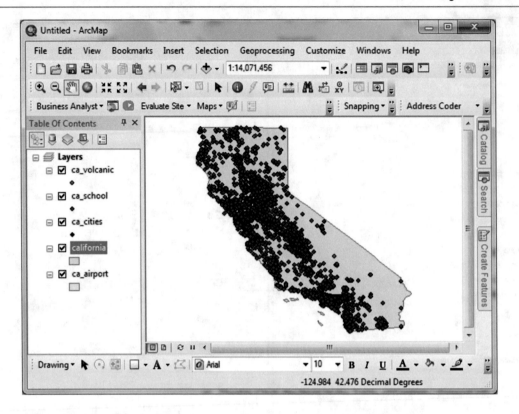

Change the Name of Layers

4. D-click "ca_volcanic" layer in the Table of Contents/General tab/Click on the Layer Name and type "Volcano" OR
5. Click twice on "ca_volcanic" and rename the file's name
6. Repeat the previous step and rename all the layers for example "ca_airport" will be "Airport", the rest "Cities", "School", "California"

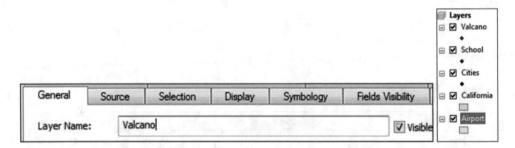

Symbolize the Layers

Symbolizing means assigning various things to the features, if you have a point feature, you can apply different colors, sizes, and shapes. If you have line feature you can apply width, and color and if you have polygon features you can apply patterns, colors, and make it transparent. Symbology is also a scale dependent. A city in California may be a polygon on one map if the scale is large such as 1:70,000 or a point on another map, if the scale is small such as 1: 4,000,000. In ArcGIS, the symbols are organized by style, such as Civic, Conservation, Crime Analysis, Environmental, Geology, and others. In addition GIS allow you to create your own symbols. The Symbology is applied to a field in the attribute table.

ArcGIS allow you to symbolize the feature and then save it as a layer, so when the layer is added to the map the feature is already symbolized the way you saved it. Symbology is also can be applied to the images. You are going to change the symbol and color of the school and the color of the airport.

(a) In ArcMap, click on the **School** symbol in the TOC
(b) The Symbol Selector display, scroll down and chose the **School 2** symbol **OR**
(c) Type School in the Search window/click search/chose the **School 2**
(d) Keep the green color and make the size 9
(e) Click "OK"
(f) R-click the symbol of the airport and change the color of the Airport to the color of your taste

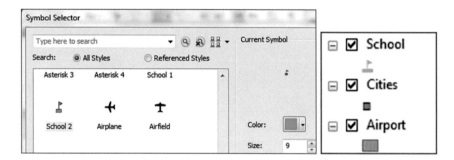

Find and Create a Bookmark

Create a Bookmark for the School North of the City of Chico

1. Click the "**find**" icon in the **Tools** toolbar

2. Under the "feature" tab:
3. Type "Chico" in the "Find" box
4. In the dropdown menu in the "In" select "Cities"
5. Click "Find" button
6. R- Click Chico in the result, select "Zoom To"

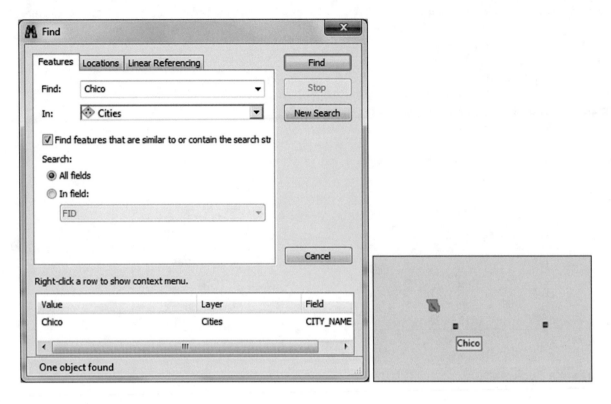

7. Close "Find Dialog"
8. R- Click on the "Cities" layer in the TOC/Properties
9. Under the "Display" tab
10. Check the box "***Show Map Tips using the display expression***"

Note: Make sure the Field: **CITY_NAME**

11. Click "OK"
12. In the Tools toolbar click on Select Elements
13. In the map put your Select Element cursor above the city that you have zoomed in

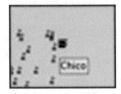

Result: The Chico city will display

14. Click the Zoom In ![icon] in the **Tools** toolbar and zoom in around city of Chico
15. Go to "Bookmarks" menu, select "Create Bookmark"
16. Give desired name (i.e. "Chico")

17. Click "OK"

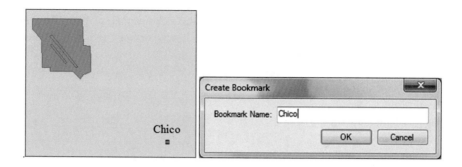

18. On the View menu click "Layout View" OR
19. Shift your work between "Data View" & "Layout View" at the bottom of the map

20. Click Data View

21. Click Full Extent button ![globe icon] at the **Tools** toolbar
22. Click " Bookmarks" and select "**Chico**"

Using Relative Paths

Relative paths in a map specify the location of the data contained in the map relative to the current location on disk of the **map document** (.mxd file) itself. As relative paths don't contain drive names, they enable the map, and its associated data, to be moved to any disk drive without the map having to be repaired. If you don't set the relative path, and you move your map document from one directory, such as (\\C:\data\), to another directory (\\K:\GIS\), (or another computer), you will encounter a problem when you display your layer data. The layer will still appear in ArcMap TOC with red exclamation mark, but will not display in the Data View. To avoid this issue, it is recommended to set the relative path.

1. Click the File menu/click Map Document Properties
2. Check the option to store relative path names and click **OK** to apply the settings.

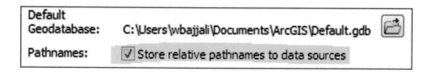

Save Map Document

Save the map document to the result folder

3. File/Save As/browse to \\Ch02\Result
4. Save it as **Chico.mxd**

   Quiz: Find Fountain Valley School and Bookmark it

Lesson 2: Data Management

Scenario 2: In this exercise, you are going to work with spatial data from the state of Texas. In the state, there is a pipeline that runs from Lubbock to San Antonio, passing various cities and transporting fresh water. The fresh water is abstracted from a deep groundwater wells. Your duty is to do the following:

1. Open new ArcMap Document and Add data into ArcMap
2. Change the name of layers
3. Change Symbols and Show Map Tip
4. Label the groundwater wells
5. Symbolize the pipeline and rivers
6. Create a layer file

GIS Approach
1. Start a new ArcMap document by doing the following
2. File/New/accept the default setting/OK
3. A dialog box display asking to "Save Changes to Untitled/No
4. Click "Add data" 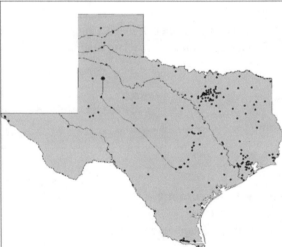 in the Standard toolbar
5. Browse to the directory \\Ch02\Data\Q2
6. Use the Shift key, highlight the following files: **gw.shp, pipeline.shp, tx.shp, tx_city.shp, and tx_river.shp**
7. Click "Add"
8. A message will be displayed stating "**Unknown Spatial Reference**"

Note: this means that the following data sources you added are missing spatial reference information (will be discussed in projection chapter)

9. Click OK

Result: the 5-layers will be displayed

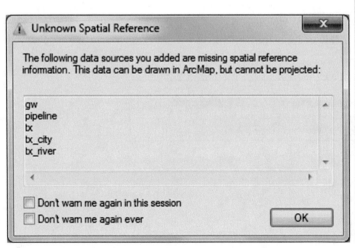

Change Names

Change the name of all layers and give them a proper name i.e. tx_city is City

10. R-click on desired file
11. Select "Properties"
12. Under the "General" tab rename the "layer name" as desired Or
13. Click twice on each layer and rename it as desired (see below)

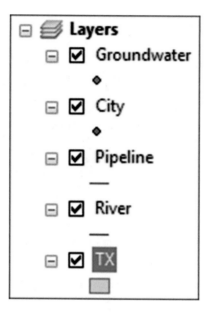

Change the Symbols of the Layers and Show Map Tip

Change the symbol of some layers and identify city names located along the pipeline using Show Map Tip

1. Click on the **City** symbol in the TOC, select Square 2, Size 7, and the Color to Ginger Pink then click OK
2. Change the symbol and color of the Groundwater into Circle 2, Size 8, and the Color to blue
3. Click the symbol of the Pipeline, type in the Symbol Selector "Pipeline" click search select "***pipeline_segment_line***", then click OK
4. Click the Symbol of the River, select the River Symbol, then click OK

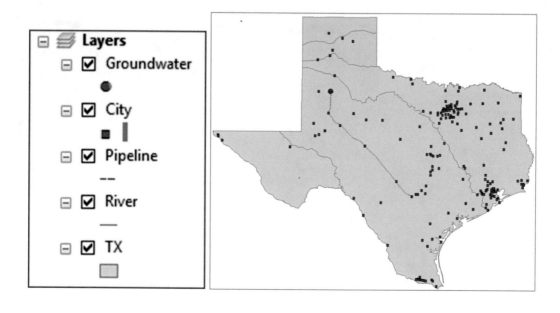

1. Uncheck the "Groundwater" layer in the TOC
2. Zoom In around the Pipeline
3. R-click on the "City" layer/Select "Properties"
4. In the "Display" tab, click the "***Show Map Tips using the display expression***" box and the Field: ***NAME***
5. Click "OK"
6. Click on the Select Element icon in the Tools toolbar

7. Place it above each city along the pipeline in order to see the name of the cities

   Result: Lubbock, Big Spring, San Angelo, Kerrville, and San Antonio

Label a Layer

There is another way to see the name of the features in the layer by using the labeling. You are going to label the Groundwater layer using the "ID" Field

  8. R-Click Groundwater layer in the TOC
  9. Select "Zoom To Layer"
10. R-Click "Groundwater" layer/Label Features
     Result: The Groundwater wells will be labeled, but not with the ID field.
11. In order to label the Groundwater layer with the ID field, do the following
12. R-Click groundwater layer once again
13. Select "properties"
14. Click the "Labels" tab check the box "label features in this layer"
15. Label Field "ID"
16. Click Symbol
17. Change the Aerial into Time New Roman, Size 12
18. Click "OK" then "OK"

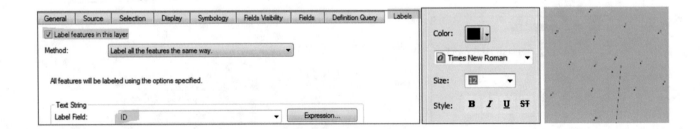

Create a Bookmark

Create a Bookmark for the Groundwater Wellfield

19. Click "Bookmarks" menu, select "Create Bookmark"
20. Give desired name (i.e. "GW")
21. Click "OK"
22. Click Full Extent
23. Click "Bookmarks" menu select "GW"
24. Click Full Extent

Classification

Classifying feature is an important part in GIS map making. Symbolizing is based on a field in the attribute table of a feature class. The attribute table of the city's layer contains a numeric attribute called "POP1990". You would use this field to symbolize the city layer.

Classify the City Layer

a. D-click the "City" layer/Symbology/Categories/
b. Unique Value
c. Value Field: **POP2007**
d. Click "Add All Values"
e. R-Click Color Ramp/Uncheck Graphic View/Choose Pastels/OK

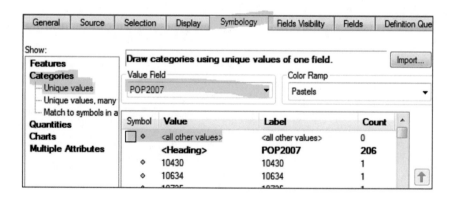

QUESTION: Does this classification make a sense?

ANSWER: NO

Note: in Chap. 3 you will learn how to perform better symbology and classification

Map Classification and Layout

3

Introduction

Map layer, that represent features in GIS, has more than its location and shape. The GIS layer can be associated with different information. For a river, this might include its name, length, and its water quality. For a county, this might include its population, ethnic group, household, income, age, and others. The information associated with a feature in a GIS is called an attribute. For example, population can be an attribute of a city, country, and other features. Feature attributes are stored in an attribute table. In an attribute table, each feature is a record and each attribute is a field. The attributes for all the features in a layer are stored in the same attribute table.

FID	Shape *	NAME	POP2004	MALES	FEMALES	AGE_5_17	AGE_18_21	AGE_22_29
0	Polygon	Bayfield	15866	7590	7423	2906	520	877
1	Polygon	Taylor	19905	9966	9714	4184	917	1622
2	Polygon	Marinette	44538	21415	21969	7979	2388	3069
3	Polygon	Langlade	21190	10291	10449	3936	833	1526
4	Polygon	Washburn	17108	8071	7965	2991	617	1029
5	Polygon	Burnett	16433	7897	7777	2700	607	988

This attribute table for a layer of county stores each feature's ID number (FID), shape, name, population in 2004, males, females, age of populations, and others.

A feature on a GIS map is linked to its record in the attribute table by a unique numerical identifier (ID). Every feature in a layer has an identifier. Because features on the map are linked to their records in the table, you can click a feature on the map, and see the attributes stored for it in the table. When you select a record in the table, the linked feature on the map is automatically selected, as well, and vice versa.

When you add data to ArcMap, a random colors for the layer symbols is assigned. You can change the colors and assign a color of your choice to make the map easy to view. When applying proper symbols and classification to the map in GIS, the map becomes easy to understand. ArcGIS offer diverse symbol and label styles that user(s) can use on maps, and can modify them so the maps look just as desired. Features can also be symbolized based on an attribute. Maps, on which features have been symbolized based on an attribute often, convey more detail and clarification. For instance, road lines could be symbolized by a type attribute to indicate different roads, such as highway, interstate, or major road. Individual well locations could be symbolized by a yield attribute, to show the capacity of wells by discharge in m^3/h.

The type of symbology depends on whether an attribute's values are text or numbers. The numbers represent counts, amounts, rates, or measures. When a layer is symbolized based on an attribute with a text value, features are represented with a different symbol. Exactly how the symbols differ from one another depends on what you are mapping. For instance, if you were symbolizing geology, according to the outcrop formations, you might use polygon symbols with different shades or

Electronic Supplementary Material: The online version of this chapter (https://doi.org/10.1007/978-3-319-61158-7_3) contains supplementary material, which is available to authorized users.

color to represent the different formations. But, if you were mapping streams according to base flow, you might show streams with permanent base flow as a solid line, and the intermittent stream, as a dashed line.

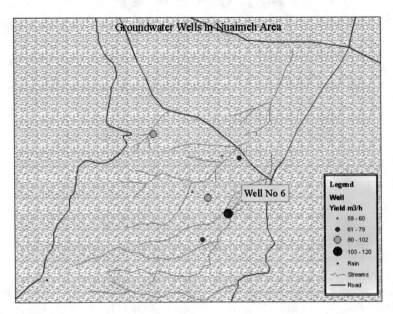

The map shows the groundwater wells in Nuaimeh area—Jordan symbolized based on their yield. The highest yield is associated ith well # 6

Feature quantities are typically represented on a map by creating groups of features with classes and assigning a special symbol to each class. The most common ways to symbolize quantities are the graduated symbols and colors.

Displaying features in a graduated sequence, allows the map to be visualized the distribution patterns in quantity data. For example, the groundwater map above is drawn with colors ranging from blue to green to red. The red wells can be interpreted to represent greater values than green and blue colors. Likewise, it is clear that blue the smaller symbols represent smaller well yield than larger symbols.

Lesson 1: Creating Map and Data Classification

Scenario: You are a geologist working for the Ministry of Water and Irrigation in the Azraq basin. You have been asked to prepare a map showing the major outcropping formation and the salinity of groundwater wells in the whole basin. You have been asked to evaluate the effect of geology and structure on the yield of groundwater.

Creating Geology Map Using Text Attribute

Data classifications are useful for representing continuous data in logically defined categories for use with mapping. The geology shapefile is a polygon layer and has a qualitative attribute called "Lithology" field. The lithology field consists of different local geological codes will be used to create a color-coded map. The map depicts the outcropping geological formations in Azraq basin.

Data Integration

1. Launch ArcMap
2. Click on Add Data ![Add Data icon] on the Standard toolbar
3. Once the dialog box pops up click "***Connect To Folder***" ![Connect To Folder icon] button.
4. Connect to \\Ch03\Data folder to create a shortcut that can be used throughout the lesson.

5. Add the following data **Fault.shp**, **Geology.shp**, and **Well.shp** by holding the control key to select and highlight multiple pieces of data at once.
6. Then click Add.

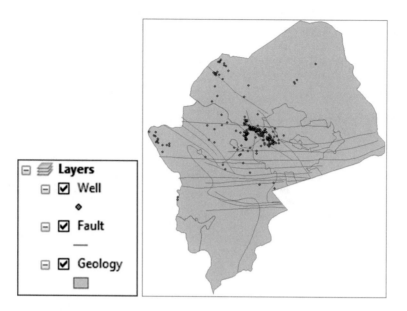

Rename the Layers Data Frame and Call It *"Azraq Basin"*
7. In the TOC, R-click on Layers/click on properties.
8. Click on the General tab.
9. In the Name box select the Layers, and click delete and type in Azraq Basin.
10. Then click OK

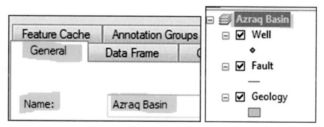

Result: the new name will appear on the data frame.

Create the Geological Map Using the Lithology Field
11. R-click on the Geology layer, click on the "Open Attribute Table"
12. The "Lithology" field will be used to symbolize the Geology layer.

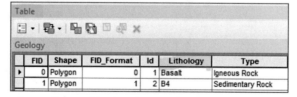

13. Close the table after looking at the attribute table, by clicking on the "X" in the right hand corner.
14. R-click on the Geology layer and go to Properties, or double click on the Geology layer and it will automatically make the Layer Properties box to appear.
15. In the Layer Properties click on the Symbology tab.
16. Click on the Categories under the "Show" box
17. Select the Unique value option
18. In the Value Field box select Lithology, and then click on the "Add All Values".

19. Uncheck "all other values"
20. Click OK
21. In the TOC, click on the B3 symbol of the Geology layer and it will take you to the Symbol Selector dialog box.
22. In the lower right hand corner click on the "Style References" and the Style References box will appear.
23. In the Style References box scroll down until you find the "Geology 24K". Select it and click OK.

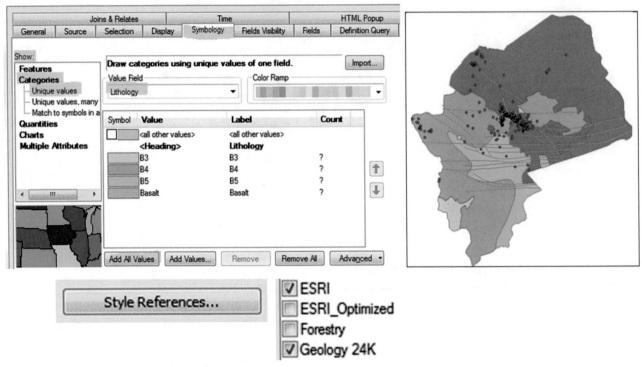

Result: in the Symbol Selector box the "Geology 24K" has been added.

24. In the Symbol Selector scroll down and select "*624 Carbonaceous Shale*", on the right hand side, click fill color and pick "*Quaternary 2*" (Column 2, Row 11) and make the outline width 0.1, and outline color "**Gray 60%**", (C1, R7), and then click OK

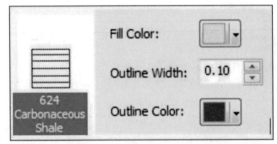

25. In the TOC click on the B4 symbol of the Geology layer, scroll down and click 627 Limestone, click fill color and pick "*Quaternary 4*" (Column 4, Row 11) and make the outline width 0.1, and outline color "**Gray 70%**" (C1, R8), and then click OK

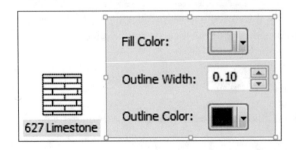

26. Repeat the Same for B5 and select "638 Argillaceous Limestone", click fill color and pick "*Tertiary 4*" (C8, R11) and make the outline width 0.1, and outline color "**Gray 80%**" (C1, R9), and then click OK

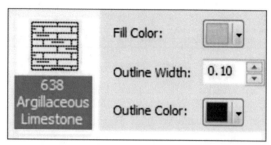

27. Click the Basalt Category, select "407 Igneous", click fill color and pick "*Volcanic 6*" (C5, R18) and make the outline width 0.1, and outline color "**Black**" (C1, R9), and then click OK.

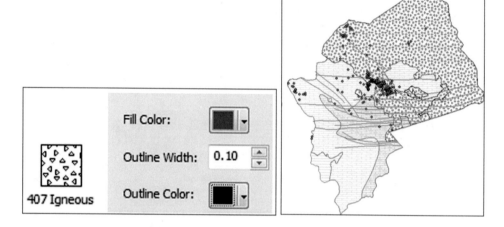

Creating Salinity Map Using Numeric Attribute

ArcMap provides many types of classification and color ramps, which can be used to highlight different aspects of the data. When classifying the data, you can use one of many standard classification methods provided in ArcMap, or you can manually define your own custom class ranges. In this section the following classification will be used

1. Natural Break
2. Quantile
3. Equal Interval
4. Manual

Classify the Salinity of Groundwater Using Natural Break Method

The Natural breaks (Jenks) classification is designed to place variable values into naturally occurring dataset. The features are divided into classes whose boundaries are set where there are relatively big differences in the data values. The set of data are classified by finding points that minimize within-class sum of squared differences, and maximizes between group sums of squared differences. The advantage of this classification is that it identifies actual classes within the dataset, which is useful to create true representations of the actual salinity of the groundwater wells.

1. In the TOC, r-click on the Well layer, "Open Attribute Table".
2. The Salinity field in the attribute table will be used to classify the Well layer.

	FID	ID	Shape	ALTITUDE	WELL_DEPTH	YIELD	S_WATER_LE	SALINITY
▶	0	1	Point	910	438	70	317	864
	1	2	Point	890	428	55	311	902
	2	3	Point	551	78	20	31	1574
	3	4	Point	551	75	12	27	1267

3. R-click the "Salinity" field and select "Sort Ascending"

Question: What is the highest and lowest salinity?

4. Close the attribute table.
5. In the TOC, r-click on the Well layer, click on the Properties, and the click on the Symbology tab.
6. Click on the Quantities then on the Graduated symbols
7. In the Field Value drop down box select the "**SALINITY**" field.
8. Under Classification change the number of Classes to 4, click on the Classify button, and change the Classification Method to Natural Breaks (Jenks).

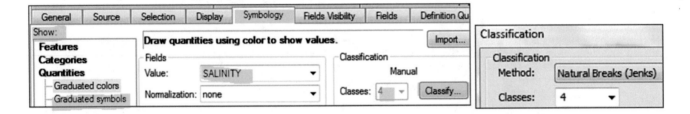

9. In the Break Values, click the first value, 1,678, to highlight it
10. At the bottom of the Classification dialog box, it shows that the class has 205 elements
11. Click the second value, 5000, it consist of 59 elements

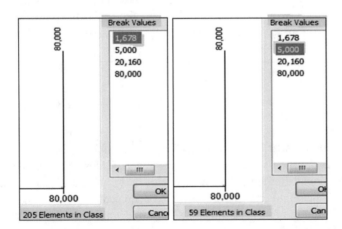

Question: how many elements in the third and fourth value?

Next you will unify the symbol shape and change the symbols sizes and colors

12. Click Symbol button under Template and choose symbol Circle 2. And then click OK

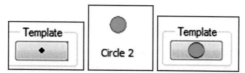

13. Click the Label heading, Format Labels, select the check box beside "Show thousands separators, then click OK

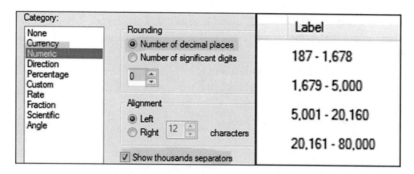

14. From the Symbol Size from, click on the smallest circle its size is 4 currently, so change it to 8 and don't change the size of the biggest circle
15. Click tab twice to exit the Symbol Size from

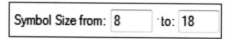

16. D-click on the biggest symbol, click color and change it to red
17. Continue by D-click on each symbol and change the color as seen below
18. Click OK to exit the Symbology tab

Result: After you have selected all of the colors and made the appropriate symbol size changes, your dialog box should look like this or very similar.

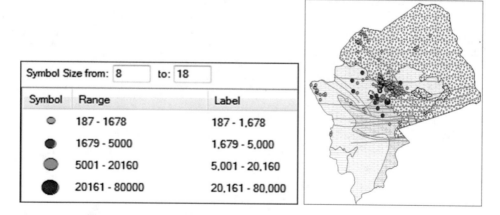

Classify the Salinity of Groundwater Using Quantile Method

In the quantile method, each class contains an equal number of features, and this method is mainly used in homogenously data. This method is useful when you want to emphasize the relative location of highly saline wells, among other low salinity wells. This method sometimes provides misleading results, because the groundwater salinity, in the wells, is grouped in equal numbers, in each class. To avoid this phenomenon, you can increase the number of classes.

19. Insert menu, click Data Frame
20. Rename the Data Frame Quantile
21. R-click the Geology layer in the Azraq Basin Data Frame and click Copy
22. R-click the Quantile Data Frame and click Paste Layer(s)
23. Repeat the above steps and copy the Well and Fault layers into the Quantile Data Frame
24. In the TOC, D-click the Well layer, and click the Symbology tab.
25. Click on the Quantities then on the Graduated symbols
26. In the Field Value make sure that the "**SALINITY**" field is selected.
27. Under Classification make sure that number of Classes is 4, click on the Classify button, and change the Classification Method to "**Quantile**".
28. In the Break Values, click the first value, 614, to highlight it
29. At the bottom of the Classification dialog box, it shows that the class has 70 elements
30. Click the second value, 958, it consist of 70 elements
31. The third and fourth numbers consist of 69 elements
32. Click OK and then OK

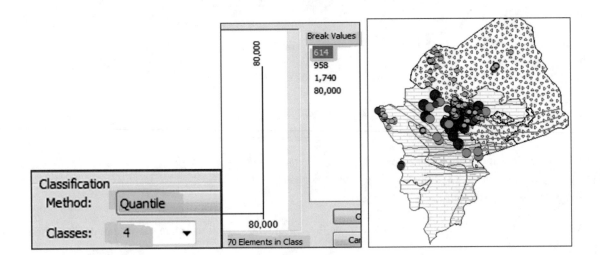

Classify the Salinity of Groundwater Using Equal Intervals Method

The equal interval method generates classes that have equal range. This is an advantage as each class will be equally represented on the map. The equal interval method is best used in recognizable data ranges, but not for heterogeneous data such in the case of groundwater salinity in Azraq basin.

33. Insert menu, click Data Frame
34. Rename the Data Frame Equal Interval
35. R-click the Geology layer in the Quantile Data Frame and click Copy
36. R-click the Equal Interval Data Frame and click Paste Layer(s)
37. Repeat the above steps and copy the Well and Fault layers into the Equal Interval Data Frame
38. In the TOC, D-click the Well layer, and click the Symbology tab.
39. Click on the Quantities then on the Graduated symbols
40. In the Field Value make sure that the "**SALINITY**" field is selected.
41. Under Classification make sure that number of Classes is 4, click on the Classify button, and change the Classification Method to "**Equal Interval**".
42. In the Break Values, click the first value, 20,140, to highlight it
43. At the bottom of the Classification dialog box, it shows that the class has 275 elements

44. Click the second, third and fourth values, it consists of 1 elements
45. Click OK

Result: The classification consists of 4 classes and the range of the 4 classes is almost identical and the first class consists of 275 wells and the rest of classes, each consists of one well only. This is shown in the symbol color which consist mainly of cyan color

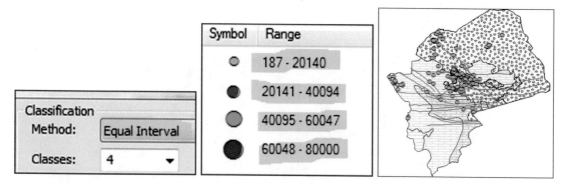

Classify the Salinity of Groundwater Using Manual Method

This method allows users to use their own classes manually, by setting the class ranges that are appropriate for the dataset. This method allows user, for example to classify the salinity based on the known drinking water quality standard or the water-rock interaction standard. The method is appropriate for classification the salinity of groundwater as it allows to emphasize features with particular values, for example, those wells that are highly saline at certain locations.

46. Insert menu, click Data Frame
47. Rename the Data Frame Manual
48. Click ctrl and then click the Geology, Well, and Fault layers in the Equal Interval Data Frame
49. R-click the Manual Data Frame and click Paste Layer(s)
50. In the TOC, D-click the Well layer, and click the Symbology tab.
51. Click on the Quantities then on the Graduated symbols
52. In the Field Value make sure you have the "**SALINITY**" field.
53. Under Classification make sure you have 4 Classes and click Classify button, and change the Classification Method to Manual.

Note: The groundwater salinity is measured in mg/l. You would like each class to represent the quality of groundwater based on the Jordanian drinking water quality standard and water rock interaction.

54. Under break values change the values as below

Old Value	New Value
1678	1000
5000	2000
20160	4000
80000	80000

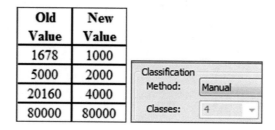

55. After you have changed everything, click on an empty place and then OK and then OK to exit the Layer Properties dialog box

Result: the changes can be seen in the Map view

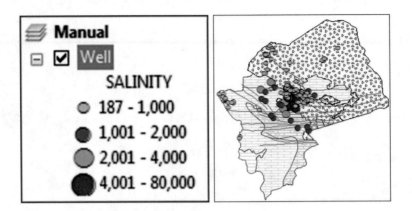

Create Map Layout for the Groundwater Classification

The layout is a collection of maps that created in the different data frame designed to be saved as an image, or used in word document, pdf file, PowerPoint or printed as a poster or simply distributed as a color-coded map. Page layouts can be a landscape or portrait orientation. To the layout map, you can add legend, scale, title, neatline, north arrow, and much more.

Scenario: Your boss asked you to create 8.5 by 11 inches layout page map with landscape orientation in pdf format. The map document will be distributed to the shareholder of Azraq Basing during the annual briefing about the groundwater quality. Your pdf map document should show the Jenks, Quantile, Equal Interval and Manual classification.

56. Click View menu, then click Layout View

Note: you can also click the Layout View button at the lower left part of the Map view

57. Click Customize menu, ArcMap Options, click Layout View tab
58. Make sure that the setting match the ArcMap Options dialog box

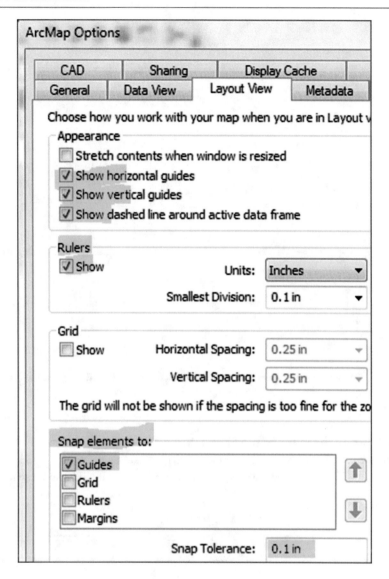

59. Click OK to accept setting
60. Click File menu, click Page and Print Setup

Note: you can also r-click anywhere inside the Layout View page and click Page and Print Setup.

61. If a printer is connected to the computer, select it
62. For paper size and page size, select "Letter" (8.5 by 11 inches)
63. For Orientation select "Landscape"
64. Check Use Printer Paper Setting
65. Check Scale Map Element proportionally to changes in Page Size
66. Click OK

Result: The layout View is displayed in landscape orientation

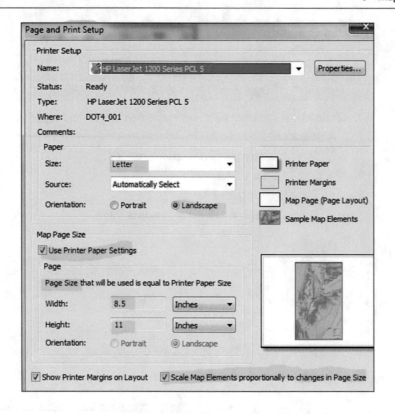

67. Click OK

Use the Guidelines to Arrange the Maps in the Layout View

68. Click at 0.3 in. on the horizontal ruler to create a vertical blue guide at this location
69. Do the same at 5.4. 5.6, and 10.7 in. on the horizontal ruler to create vertical guides at these locations

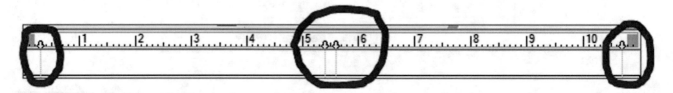

70. Click at 0.2 in. on the vertical ruler to create a horizontal blue guide at this location
71. Do the same at 4.0. 4.2, and 8.2 in. on the vertical ruler to create horizontal guides at these locations

Result: The vertical and horizontal guides are now seen and will be used to snap the four maps in the layout view

72. Click the Azraq Basin map (Jenks) in the Layout to select it (once it selected the Azraq Basin Data Frame in TOC will be activated)
73. Drag it so its upper left corner to snap to the intersection of 0.3 in. on horizontal and 8.2 in. on vertical rulers respectively
74. Drag the lower right corner to snap to the intersection of 5.4 in. on horizontal and 4.2 in. on vertical rulers respectively

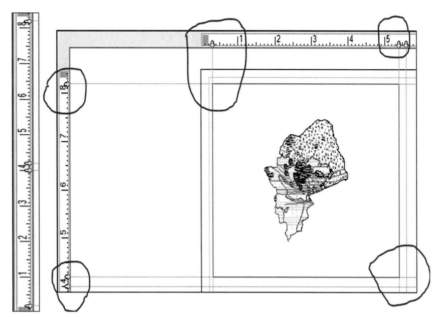

75. Repeat the steps above and drag the following to the following guides in the horizontal and vertical rulers (see table below)

Map	Upper Left Corner			Lower Right Corner	
	Horizontal	Vertical		Horizontal	Vertical
Quantile	5.6	8.2		10.7	4.2
Equal Interval	0.3	4		5.4	0.2
Manual	5.6	4		10.7	0.2

76. Select the Jenks Classification map and click Full Extent ⊙ button
77. Repeat this for the Quantile, Equal Interval, and Manual map respectively

Insert Title
78. Select the Jenks Classification map, click Insert menu, Title
79. D-click the title and type "**Jenks Classification**" in the Insert Title text box
80. Click Change Symbol, select New Roman Times and size 18, and click Bold (Use Draw toolbar)
81. Click OK/OK
82. Repeat the above steps and insert title and type "Quantile Classification", "Equal Interval Classification", and "Manual Classification" in the Quantile, Equal Interval, and the Manual map respectively

83. Use the Pan 🖐 button to move each map in the layout to the right to accommodate the title in the upper left corner of each map (see below)

Insert Legend
84. Click the "Jenks Classification" map element in the layout to activate its frame
85. Click Insert menu and click Legend
86. Make sure that the Well, Fault, and Geology layers are under the Legend Items
87. Click Next
88. Under Legend Title, delete Legend and type "Salinity Classification"
89. Size 16 and font "Time New Roman", click Next

90. In the Border click Style Selector  and select 1.0 point
91. Click OK, Next, Next, and Finish
92. Zoom to the upper left side of the "Jenks Classification" map element
93. Click at 0.5 and 2 in. on the horizontal ruler to create a vertical blue guide at this location and 4.3 in. on vertical ruler to create horizontal blue guide.
94. Drag the Legend so that it snap to the 0.5 and 2 in. horizontal guide and 4.3 in. vertical guide
95. Repeat the above steps and insert Legend for the "Quantile Classification", "Equal Interval Classification", and "Manual Classification" respectively (see the image below for a suggestion location).

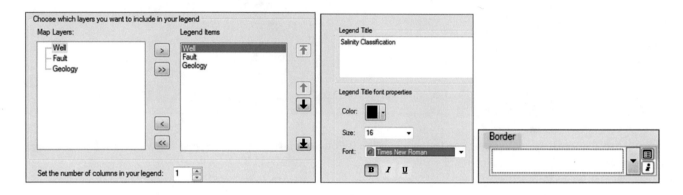

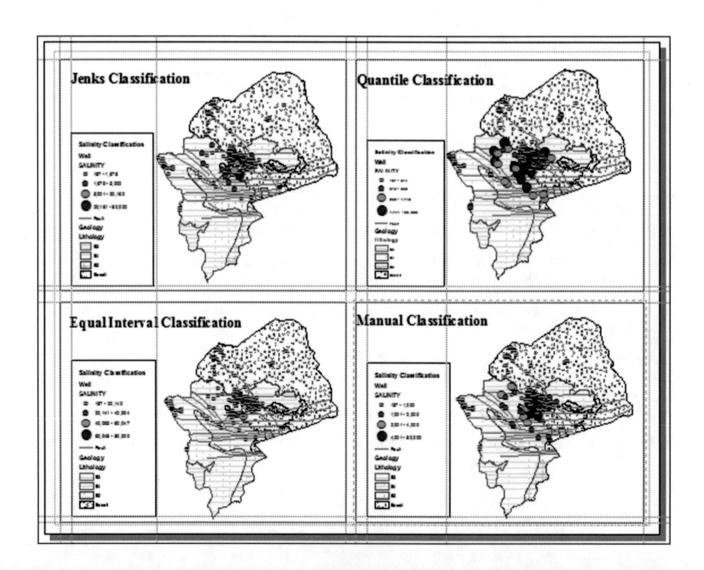

Insert Scale Bar and North Arrow

96. Click the "Jenks Classification" map element in the layout to activate its frame
97. Click Insert menu and click Scale Bar
98. Click Alternating Scale Bar 2

99. Click Properties
100. Click When resizing drop-down arrow and click "Adjust width"
101. Division Units: Kilometers
102. Label Position: below center
103. Division value: 40 km
104. Number of divisions: 2
105. Number of subdivisions: 1
106. OK
107. Drag the scale to lower right corner

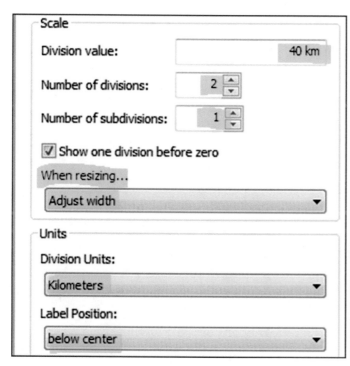

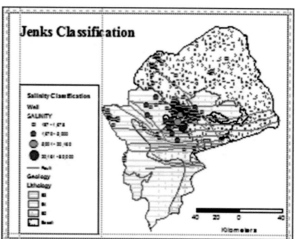

108. Insert menu and click North Arrow, click ESRI North 1
109. Drag the north arrow above the scale

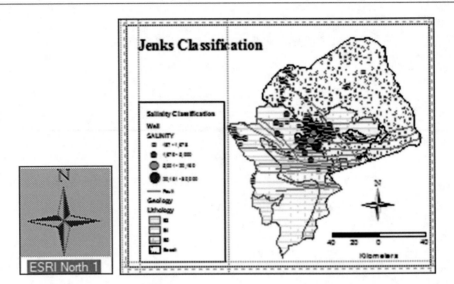

110. Repeat the above steps and insert Scale Bar 2 and North Arrow to the "Quantile Classification", "Equal Interval Classification", and "Manual Classification" respectively

Data Acquisition and Getting Data into GIS

Section I: Introduction to the Acquisition Method

Data entry process is the most important step and the most time-consuming part of the GIS operation. User should always be aware in advance at what type of data is needed for the project and where to locate or obtain it. Reliable and accurate data is essential for carrying meaningful GIS project. GIS users can rely on many sources of GIS data and some of the data is already available in different format or it can be created. Some of the most common source of GIS projects are hard copy maps. The maps can be digitized or scanned and integrated into GIS as a digital data. There are also huge digital spatial data, available from various organizations and commercial sources. They are processed data of different types and ready to use. Some of the data is free, you can download it, copy it, and borrow it. You can even convert it from one format to another.

Data is also available in tabular format. It can then be integrated into GIS one of two ways. Tabular data that has fields with a coordinate system can be integrated directly. Data that lack the coordinate system can be joined or linked into GIS, then the data can be converted into a layer (Discussed later). Reports containing relative and important data for a specific project can be translated into useful data by manual entry or scanning them. Then stored as an associated ingredient of the database. Reports can be any valuable information that can be turned into GIS layers, such as classification or even explanation for a certain type of data. Field Data is considered a primary data that a user can collect and integrate into GIS as a database. User can use GPS to register the location in term of X, Y coordinates and altitude of the observation locations image below. The data that is gathered in the field can be joined to the GPS file and integrated directly into GIS.

Images are also major source of GIS data. Imagery can be placed directly into the GIS database. Remote sensing is collecting data of a landscape from above, such as an aircraft that has cameras and electronic sensors.

Remote Sensing has several advantages that are important in GIS. Satellite imagery is very consistent regarding data quality and the condition of collection. They cover a very large area and provide a permanent record that can be verified and used for a long time. This will allow keeping records that show the changes that took place due to a certain activities.

This chapter will provide the following exercises:

1. Integrate flat file data into ArcMap
2. GPS Data Integration into ArcMap
3. Data Integration from the Internet
4. Add Data from ArcGIS Online
5. On Screen Digitizing

Electronic Supplementary Material: The online version of this chapter (https://doi.org/10.1007/978-3-319-61158-7_4) contains supplementary material, which is available to authorized users.

W. Bajjali, *ArcGIS for Environmental and Water Issues*, Springer Textbooks in Earth Sciences, Geography and Environment, https://doi.org/10.1007/978-3-319-61158-7_4

Field collection using GPS

Table 4.1   Coordinate in latitude and longitude

Proposed site	Longitude			Latitude		
WWTP	Degree	Minute	Second	Degree	Minute	Second
1	91	40	42	46	27	11
2	92	8	50	46	40	15
3	92	2	6	46	28	16

Integrate Flat File Data into ArcGIS

Scenario 1: You are a professional working for the city of Superior, WI. Your supervisor gave you a word report that has a table containing three proposed locations of a Waste Water Treatment Plant (WWTP) in Douglas County (Table 4.1). The coordinate is in degree, minute, and seconds (DMS) and your duty is to do the following:

1. Use the Excel software to convert the DMS into Decimal Degree (DD)
2. Save the file as text (tab delimited) (".txt") or (comma delimited) (*.csv).
3. Integrate the text file into ArcMap
4. Convert the file into shapefile.

Convert Degree, Minute, Second into Decimal Degree Using Excel

1. Highlight the data in Table 4.1/copy it (Ctrl C)
2. Open Excel and place your cursor on A1 paste (Ctrl V) (or type the table in Excel)
3. Type "Longitude" in cell I2 and "Latitude" in cell J2
4. Place your cursor in cell I3 and type "=" then "-" ,open a bracket move your cursor to B3 cell type "+" and type "/" and type 60, type "+" and move your cursor to D3 cell and type "/" and type 3600, close the bracket and enter

=-(B3+C3/60+D3/3600) this is DMS longitude formula

$$-\left(D+\frac{M}{60}+\frac{S}{3600}\right) = -\left(91+\frac{40}{60}+\frac{42}{3600}\right) = -91.6783$$

Note: the sign '−' means west such as the location of USA (North hemisphere and second quarter)

5. Repeat these steps for the latitude coordinates using the same formula but without the '−' sign and the bracket =E3+F3/60+G3/3600 this is DMS latitude formula

$$D + \frac{M}{60} + \frac{S}{3600} = 46 + \frac{27}{60} + \frac{11}{3600} = 46.45306$$

J3		⋮ ✗ ✓ f_x	=E3+F3/60+G3/3600							
	A	B	C	D	E	F	G	H	I	J
1	Proposed Site	Longitude			Latitude					
2	WWTP	degree	minute	second	degree	minute	second		Longitude	Latitude
3	1	91	40	42	46	27	11		-91.6783	46.45306
4	2	92	8	50	46	40	15			
5	3	92	2	6	46	28	16			

6. Highlight the cell I3, the calculated longitude (−91.6783), and drag down the dot in the bottom right corner of the cell to the row of the last coordinate in the table
7. Repeat these steps for the latitude coordinates

Longitude	Latitude
-91.6783	46.45306

Longitude	Latitude
-91.6783	46.45306
-92.1472	46.67083
-92.035	46.47111

8. Highlight I3:J5 cells of the calculated longitude-latitude and copy (Ctrl C),
9. Then, R-click the cell I3 and select "Paste Special" and select "Values".
10. Then delete Row1 and the B, C, D, E, F, G, and H columns and remove the border of column A

	I	J
1		
2	Longitude	Latitude
3	-91.6783	46.45306
4	-92.1472	46.67083
5	-92.035	46.47111

Paste Special

Paste

○ All
○ Formulas
◉ Values

	A	B	C
1	WWTP	Longitude	Latitude
2	1	-91.6783	46.45306
3	2	-92.1472	46.67083
4	3	-92.035	46.47111

11. Click File menu/Save As/browse to \\Result folder, change Save as type to Text (Tab delimited) (*.txt) and call the file **Coord.txt**

File name: Coord.txt

Save as type: Text (Tab delimited) (*.txt)

Integrate the Data Table into GIS

12. Launch ArcMap.
13. Click on Add Data browse to \\Ch04\Data\Q1 folder, highlight Douglas and click Add.
14. Click on Add Data again and browse to \\Result folder, highlight **Coord.txt** (created in the earlier step) and Add.
    Note: You can also access the file in "TabDelimited" folder in Chap. 4
15. In the TOC/r-click the **Coord.txt** and go to Display XY Data.
16. Make sure the X field is the Longitude, the Y field is Latitude, and the Geographic Coordinate System is GCS_North_American_1983
17. Click OK/OK

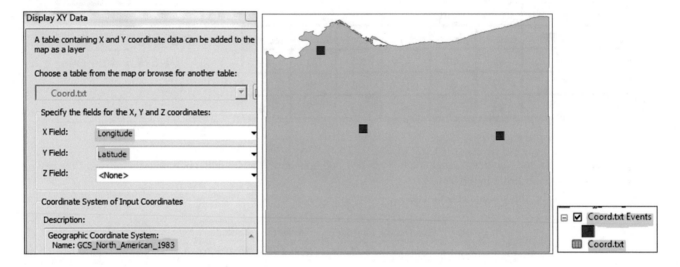

Result: Three features added into ArcMap and a layer called Coord.txt Event added to the TOC.

Note: XY event data sources are commonly used to draw point data originating from a data source that is not spatially enabled. In this respect, XY event data sources are a powerful way to integrate simple point data into your map. The Coord. txt Event can be converted into a spatial data source such as a feature class in geodatabase or shapefile.

18. R-click on the **Coord.txt Events**/Data/Export Data
19. Save it in \\Result folder and call it **WWTP.shp**
20. Click OK to add the **WWTP.shp** to the TOC
21. In the TOC/R-click on the **Coord.txt Events** and click remove

Global Positioning System and GIS

The Global Positioning System (GPS) plays a key role around the world as part of the global information infrastructure and takes serious responsibility to provide the best possible service to civil and commercial users worldwide. This is as true in times of conflict as it is in times of peace. GPS is a worldwide radio-navigation system formed from a constellation of around 30 satellites and their ground stations. GPS uses these "man-made stars" as reference points to calculate positions accurately to a matter of meters. In fact, with advanced forms of GPS you can make measurements to better than a centimeter. GPS is used in GIS extensively and the captured data can be integrated directly into GIS.

GPS Data Integration into ArcMap

If you use your GPS in the field to capture waypoints data such as well locations, or track a creek, you can get those features into ArcMap for further analysis directly. The two simplest ways for integrating GPS data into ArcMap is using either ArcGIS Explorer or the Desktop DNR Garmin Extension if you have Garmin GPS.

GPS Data Integration Using DNR Garmin Extension

Department of Natural Resources (DNR) provides an extension that will give the GPS user the ability to directly transfer data between Garmin GPS handheld receivers and ArcGIS software packages. Using this extension a user can use point features in a shapefile format and upload them to the GIS as Waypoints. The extension and documents can be downloaded free of charge from the following web page:

http://www.dnr.state.mn.us/mis/gis/DNRGPS/DNRGPS.html

Download Program

- **UPDATED** DNRGPS 6.1.0.6 RELEASES (6/2/2014)

  - DNRGPS for ArcMap 10.0 (40MB)

  - DNRGPS for ArcMap 10.1 (40MB)

  - DNRGPS for ArcMap 10.2 (40MB)

Download GPS Data and Integrate the Waypoints into ArcMap

Scenario 2: You are a researcher working for Douglas County and you have been asked to capture some GPS location at the territory of UWS campus to build a culvert to drain the surface runoff. You used the Garmin GPS and captured 12 locations. At your office, you connected your GPS to your desktop and downloaded the captured GPS waypoints.

1. Click the DNRGPS for ArcMap 10.2 (40MB) and download the program in a "GPS" folder
2. Unzip the "**dnrgps_6_1_0_6_for_ArcMap_10_2.zip**" in new folder called "**DNRGPS**"
3. Connect your GPS to your PC or Laptop to where you want to download your GPS data and switch on your GPS Garmin device (**DNRGPS program in Ch04**)
4. Open the "**DNRGPS**" in \\Ch04 folder and D-click **dnrgps.exe** 📱 dnrgps.exe
5. Click Run to excecute the program
6. The DNR GPS program opens, if the program doesn't recognize your GPS device, click GPS menu and point Find GPS

Result: The GPS device will be recognized and its name (garmin-e trex Legend) will be displayed on top

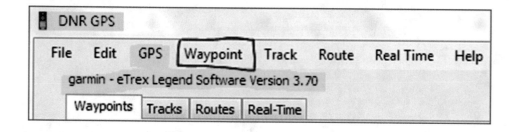

7. Click Waypoint menu and point to Download
8. 12-Waypoints will be downloaded

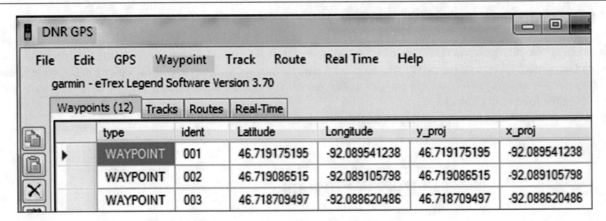

Save the downloaded Waypoints as Shapefile

9. In the DNR GPS/click File menu/Save To/ArcMap/File
   a. File Name: Culvert
   b. Save as type: ESRI Shapefile (2D) (*.shp)
10. Click Save

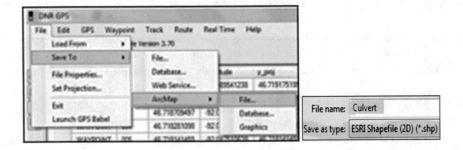

11. Make ArcMap active
12. Click Insert menu and Add data frame and call it "GPS"
13. Click Add Data browse to \\Data\Q1 folder and add Tract.shp
14. Click Add Data browse to \\Result folder and add Culvert.shp
    Note: You can also access Culvert.shp in the "Culvert" folder in Chap. 4

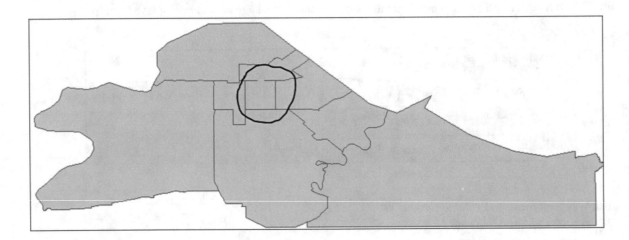

15. Save your map as GPS.mxd and exit ArcMap

Add GPS Data to ArcMap Using ArcGIS Explorer Desktop

The majority of GPS devices can export the collected waypoints as an exchange format (GPX). A format just about all GPS devices support.

16. Launch ArcGIS Explorer Desktop

17. Click Add Content button/GPS Data Files
18. Browse to \\Data\Q1 folder highlight "**Waypoint.gpx**" and click Open

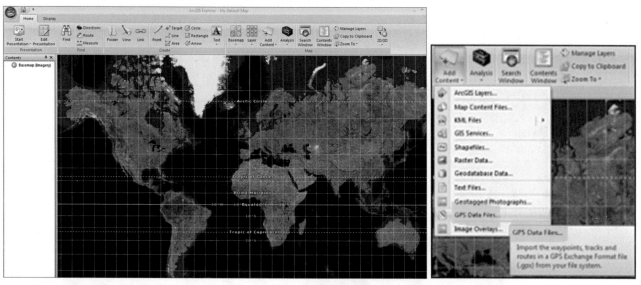

19. In the Add GPS Data dialog box/check Waypoints and click ADD

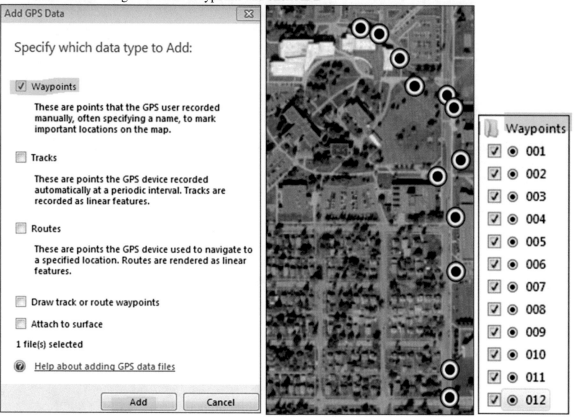

Result: the GPX file (waypoints) is added to TOC and displayed in ArcGIS Explorer Desktop with all the 12 symbols.

Save the Waypoints Layer as a Layer Package (LPK)

ArcMap support the Layer Package (LPK) and once it shared it will capture the symbols of the Waypoints for display in ArcMap.

20. In TOC/r-click the Waypoints layer and choose Share.
21. In the Sharing dialog box/highlight "Layer Package"/Next
22. Browse to \\Ch04\Output folder and save it as Waypoints.lpk

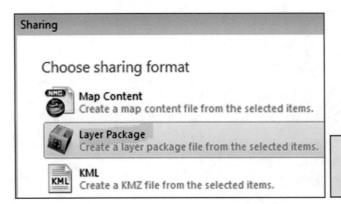
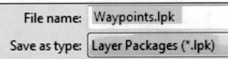

23. In the Package Properties dialog box enter "Waypoints" as the name of the package
24. Enter the description for the package "GPS waypoints" and Click OK
25. Close ArcGIS Explorer Desktop

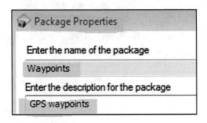

Integrate LPK into ArcMap

26. Launch ArcMap/click Add Data and browse to \\Data\Q1 folder and add Tract.shp
27. In Ch04/open the folder "**Output**" and d-click the "**Waypoints.lpk**"

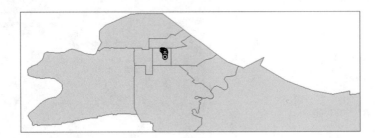

Result: The Waypoints displayed in ArcMap

Data Integration from the Internet

There are large sets of data available in either raster or vector format, which can be downloaded for free from the internet or ESRI. The data can be downloaded and integrated directly into ArcGIS and then can be used for analysis.

Note: If you have difficulty downloading the data, they will be available for you in chapter 4 folder of "Download".

Scenario 3: You have been asked to download the census tract of Wisconsin lakes and rivers in digital format for Douglas County, Wisconsin from the US Census Bureau web page.

1. Go to "www.census.gov" web page
2. Click "**Geography Map, Products**" tab

3. Click Map & Data

4. Click Map & Data
5. Under Geographic Data, click on **TIGER Products**
6. Click on Tiger/Line Shapefile—New 2016 Shapefiles
7. Click on **2016** and then click on **Download** DSA
8. Then click on Web interface.

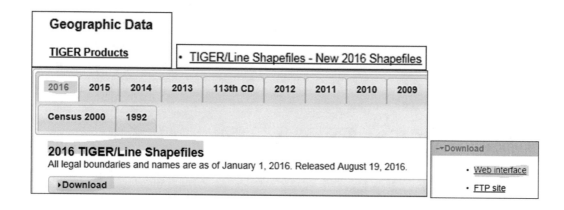

First Download the Census Tract

  9. Select year "**2016**" Select a layer type "**Census Tracts**"
10. Click Submit
11. Select **Wisconsin** as the State and click **Download**
12. Click **Save As**
13. Save the file in the \\Download folder.

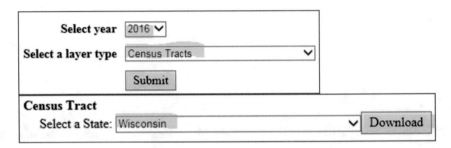

Result: The "**tl_2016_55_tract.zip**" is downloaded, now use any free software (PeaZip or 7-Zip file manager) from the internet to unzip the downloaded file.

14. R-click the "**tl_2016_55_tract.zip**" and unzip it in a folder called "**Tract**"

Second Download the Water Files

You will now download the Hydrography for Douglas County, Wisconsin

15. Select year "**2016**" Select a layer type "**Water**"
16. Click Submit
17. Area Hydrography/Select a State **Wisconsin**/Select a County Douglas County and click **Download**
18. Click **Save As** in **Download** folder as **tl_2016_55031_areawater.zip**
19. R-click the "**tl_2016_55031_areawater.zip**" and unzip it in a folder called "**Lake**"
20. Linear Hydrography/Select a State **Wisconsin**/Select a County Douglas County and click **Download**
21. Click **Save As** in **Download** folder as **tl_2016_55031_linearwater.zip**
22. R-click the "**tl_2016_55031_linearwater.zip**" and unzip it in a folder called "**Stream**"

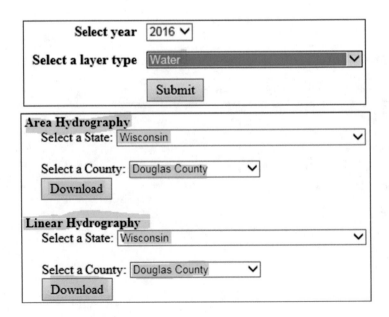

Integrate Downloaded Data into ArcMap

23. Click Insert menu/Data Frame and call it "Water Resources"
24. Add Data/browse to \\Data\Q2 folder and add **Douglas.shp**, **Lakes.shp**, and **Streams.shp**
25. Make the Douglas layer hollow, and change the streams and lakes symbols to blue

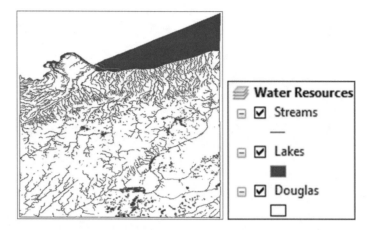

Aerial Photography and Satellite Images

Remotely sensed imagery is used extensively in conjunction with GIS technology to answer many questions about the environment in which we live. The advancement of the technology and wealth of image of various resolutions have changed the way geographic analyses are done. Images of aerial photographs and remote sensing are easily integrated into GIS. The process of integration is straightforward; images may be scanned or downloaded from the internet. The images then can be used as a background in the GIS project or can be used to be a source of obtaining vector features from the image. There are many web pages in the internet that offer free images from all over the USA.

The following web page will be used to download data from Wisconsin:

http://relief.ersc.wisc.edu/wisconsinview/form.php

The web page "WisconsinView" provides access to photographs, images, and related data regarding Wisconsin. Photographs from the National Agriculture Imagery Program; Department of Natural Resources Digital Orthophotos; and Landsat MSS, TM, and ETM+ imagery. Tornado data for Wisconsin is also provided. This web page requires you to register, which is free.

Download Image from the Internet

1. Go to the internet and type in http://relief.ersc.wisc.edu/wisconsinview/form.php

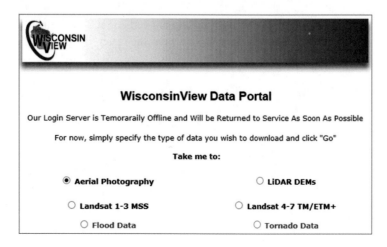

2. First enter your email address to set up your account or if you have already done this just sign in
3. Click **Aerial Photography/**then click GO.
4. Highlight the NAIP County Mosaics and year "2015" and check "UTM Version"
5. Scroll down and click on "Douglas".

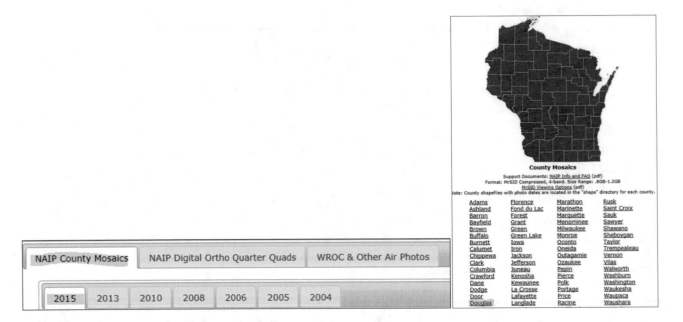

6. Scroll down and click on "**ortho 1-1 1n s wi031 2015 1.sid**"
7. Save as in \\Ch04\Download\Ortho folder

```
08/01/2016 12:00AM                  72 ortho 1-1 1n s wi031 2015 1.sdw
08/01/2016 12:00AM         995,277,169 ortho 1-1 1n s wi031 2015 1.sid
08/01/2016 12:00AM              17,004 ortho 1-1 1n s wi031 2015 1.sid.txt
08/01/2016 12:00AM              14,686 ortho 1-1 1n s wi031 2015 1.sid.xml
08/01/2016 12:00AM               1,614 ortho 1-1 1n s wi031 2015 1.txt
08/01/2016 12:00AM           Directory shape
```

8. Make sure that the Water Resources data frame is active in ArcMap
9. Add Data and browse to \\Data\Q2 folder and add "Ortho_Douglas_2015_1.sid"

Result: the "Ortho_Douglas_2015_1.sid" is added to ArcMap and aligned correctly with the three layers (Douglas, Lakes, and Streams) in the Water Resources data frame. The ortho image is the raster file that downloaded in the previous step.

10. Zoom in around Lake Nebagamon in east central Douglas County

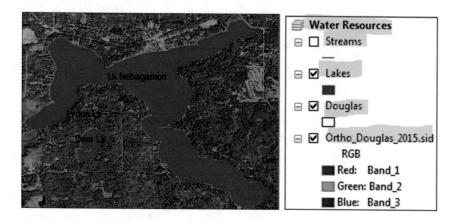

Add Data from ArcGIS Online

ArcGIS online contains set of database online of different GIS data sources. ArcGIS allows users to connect to any source desired and use it in ArcMap.

1. Insert Data Frame and call it "Protected Area".
2. Click on the arrow of the Add Data icon and click on Add Data from ArcGIS Online
3. The ArcGIS Online window will pop-up. Search for the USA Protected Area and click the search symbol.

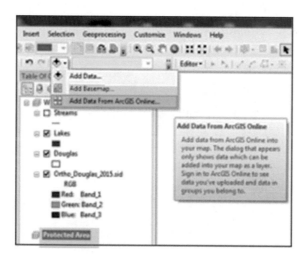

4. Click "Protected USA" and click Add.

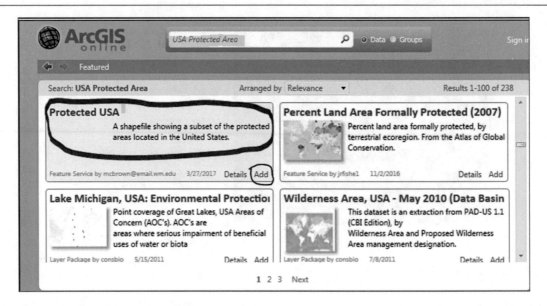

Result: The areas of concern in the USA is displayed.

5. Add Data browse to \\Data\Q3 folder and add NewtonCreek and Watershed_Newton layers
6. R-click Watershed_Newton and Zoom To Layer

Result: The Watershed_Newton and NewtonCreek displayed. The Newton Creek baseflow is treated wastewater from Murphy Oil that eventually discharge in Hug Island. The area was highly contaminated by hydrocarbon and was considered an area of concern.

7. In the Tools toolbar click Fixed Zoom Out ⬚ button several time

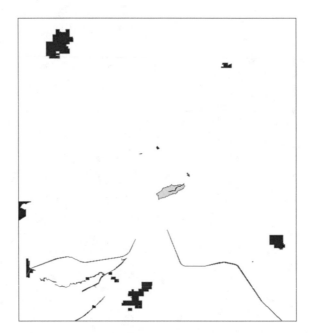

Result: Various Areas of Concern are displayed in Douglas County, where the Newton Creek was one of the sites

8. Explore other areas of concern in the USA
9. Close ArcMap and save your document as **AreaConcern.mxd** in \\Result folder

Section II: Introduction to Feature Creating

This section discusses several ways to create GIS data. GIS features can be created by converting features from existing data or creating a new data by drawing geographic features (point, line, and polygon) and adding an attribute table. Traditionally, most geographic features in GIS have been digitized from a preexisting map, scanned map, aerial photograph, or satellite photographs. Digitizing means simply capturing an analog signal in digital form or the process of tracing map features for conversion into a digital format. Digitizer allows users to transform spatial data of various types into digital format. Digitizing is a time consuming and costly process.

In this section we are going to discuss an example on screen digitizing. ArcGIS offers digitizing capability on the screen to digitize using a base layer such as an aerial photo, remote sensing imagery, or a scanned hardcopy map for visual perspective. Screen digitizing has become the most popular manual digitizing method. The feature types that can be digitized are point, line, or polygon. The features you decide to create should be identified and given a layer name prior to digitizing. The layers can be traced using the mouse and the advantage of this type of digitizing is the ease of zooming in and out for detail tracing. During the process you can use full array of editing functions such as delete, copy, move, add, and other functions.

Creating a GIS Layer from Existing Feature

This lesson shows you how you can create a new layer from an existing layer. By doing this step, the new layer will inherit the attribute table and the coordinate system from the original layer.

Scenario 4: You are a hydrogeologist and you would like to study the quality of water in the southern catchment area of the Nuaimeh region in Jordan. You have decided to create a new well layer from the southern wells in the catchment area.

GIS Approach
1. Start ArcMap and call the Layers data frame "Create Feature".
2. Click Add Data button and browse to \\Data\Q4 and highlight the **Catch.shp** and **Well.Shp** layers and click Add.
3. A dialog box display stating "Unknown Spatial Reference" (we will learn about spatial reference in Chap. 5) will appear.
4. Click Ok to display both files in ArcMap.

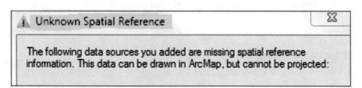

5. In TOC/click on the Well symbol, select circle 2, size 10, and blue color.
6. In TOC/R-click on the **Well** layer and click Label Features.

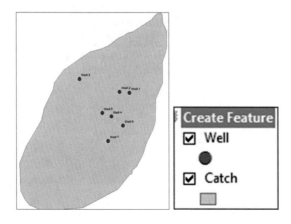

7. Go to the TOC/click "**List By Selection**" 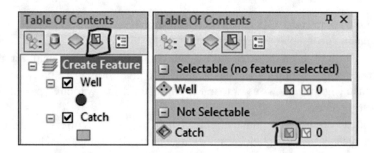 button.
8. Next, click on the **Catch** to make it "Not Selectable".

Note: This step is to select number of wells features located inside the catchment area

9. Click the drop down arrow for Select Feature tool and click on the "**Select Feature By Rectangle**".
10. Click inside ArcMap and draw a rectangle around the wells 4, 5, 6, and 7.

Result: The 4-wells will be selected and they will be seen in the TOC

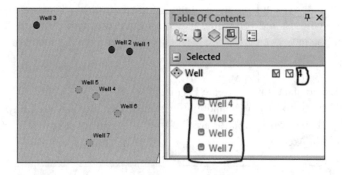

11. In the TOC/click on the "**List By Drawing Order**" 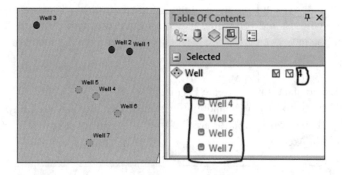 button.
12. In the TOC/r-click on the Well layer/click Data/Export Data/browse to \\Result folder and call it **Well_South.shp** and click Save.
13. A box will appear asking if you want to add the new file to the TOC/click Yes.
14. In TOC/click the **Well.shp** to uncheck it.

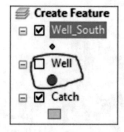

Add New Features to the Existing Layer

This lesson shows you how you can create new features in the existing layer. By doing this step, the new features will be added as new records in the well attribute table and will be displayed in ArcMap.

Scenario 5: The hydrogeological investigation showed that in the southern part of the catchment area there is a good aquifer. You have been asked to drill another four wells close to each other to generate a well-field that can be used for water supply.

1. Click Insert menu/Data Frame and rename the New Data Frame "**New Wells**"
2. Drag Catch layer from Create Feature data frame and place it in New Wells data frame. Repeat and drag the Well-South also to the New Wells data frame
3. Add Data and browse to \\Data\Q4 folder highlight "Study_Area.shp"/click Add/OK

Result: The catchment area, southern wells, and the proposed study area are displayed in the data view in ArcMap.

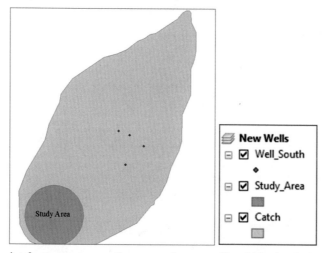

Now, you will create new four point features representing groundwater wells within the designated study area.

4. R-click on the Well_South.shp/click Edit Features/click on Start Editing.
5. The Create Feature dialog box will appear in ArcMap in the right side/click on it.

6. Click on "**Well_South.shp**" under Create Features

7. Next click on "**Point**" under the Construction tools in the Create Features Window.
8. Place a point in the Study Area by clicking on the map. Click three different places to create the three locations needed (see image below).
9. R-click "**Well_South.shp**" in the TOC/click on the Attribute Table to open it.
10. In the "Id" field type 1, 2, 3, and 4. Then under the number field in the attribute table type Well 8, Well 9, Well 10, and Well 11.

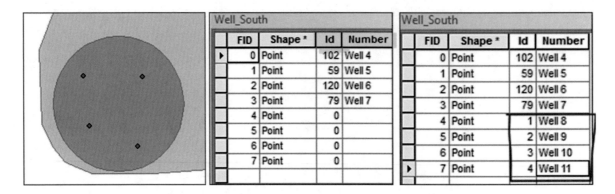

11. Once these changes have been made, close the attribute table.
12. In the Editor Toolbar click the Editor drop down arrow and click the Stop Editing

13. A box will appear asking if you want to save the edits, click Yes.

Result: The 4-new wells are now part of the Well_South layer

14. Save your map as Nuaimeh.mxd and exit ArcMap.

Digitizing on Screen

On-screen digitizing is a process in which a map is created using another map. This map could be an image, a scanned picture, or a previously digitized map. This technique is used to trace features to create new layers. This practice is similar to the traditional tablet digitizing, but rather than using a classical digitizer and a puck, the user creates layers on the computer screen with the mouse and referenced information as a background. On-screen digitizing may also be used in an editing session where the user can update or add new features. The accuracy of the digitized features cannot in any way be higher than the original base image. For accurate tracing, during digitizing, the user should zoom in for better viewing. Nevertheless, this does not mean the new captured feature will be more closely match the real world coordinate.

In this exercise you are going to use a geometrically corrected aerial photograph (orthoimage) as backdrop to capture and create new data. You will digitize on screen points (trees), sidewalk (lines), and building (polygons) from the campus of the university of Wisconsin-Superior (UWS).

Scenario 6: You are a GIS technician working for UWS and you have been asked to use an aerial photograph as a backdrop to digitize all the features on the campus in order to create an up-to-date map of campus.

Create New Polygon Shapefile

This step will allow you to add the background image and use it to trace the building in the campus of UWS.

1. Start ArcMap and rename the Layer Data Frame "**Digitizing**"
2. Click on Add Data and browse to \\Data\Q5\Image folder highlight **UWS_UTM15.tif** and click Add

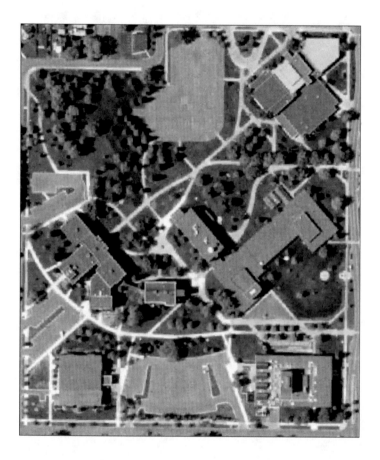

Result: the **UWS_UTM15.tif** is added to ArcMap and in the image you can see buildings (polygons), line (sidewalk) and points (trees).

3. Click Catalog window on the right hand side of the screen and browse to \\Data\Q5\folder/R-click Campus folder/New/Shapefile.
4. Name the new shapefile Building and make the feature type a Polygon

5. Click on Edit/click Add Coordinate System drop-down arrow ![icon] click Import browse to \\Q5\Image\select **UWS_UTM15.tif** and click Add/OK

Result: the new **Building.shp** is added to the TOC.

6. Repeat the step above, but make a line shapefile and name it **Sidewalk** and then create a point shapefile and name it **Tree**.

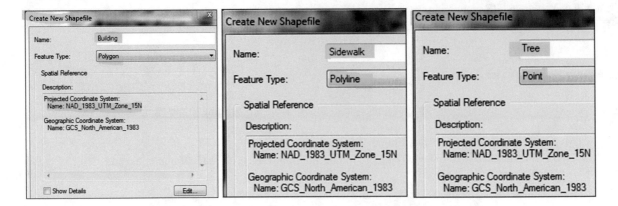

Create the Polygon Shapefile

Now you are going to start digitizing (beginning with the buildings). In order to start digitizing, you have to be in editing mode. Therefore, you have to use the Editor Toolbar and start editing.

7. In the TOC/R-click on the **Buildings** layer, click on **Edit Features** and then click on **Start Editing**.
8. Click on the **Create Features** window and then click on **Buildings**, and then "**Polygon**" under **Construction Tools**.
9. Zoom in to the lower right building in the image (the corners are labeled from 1 to 9)

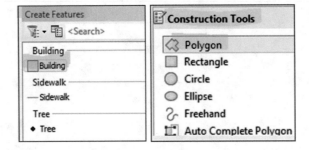

10. Position the crosshair cursor on the lower right corner of the building (No 1) to place a vertex and then move your pointer up to the north (No 2), then click again. You now have created two vertices with a straight line connecting them to define the eastern boundary of the building.

11. Mover your pointer and click a series of vertices along the perimeter of the building one at a time to form the polygon and double-click to place the last vertex or click F2.

Note: if you make a mistake and want to cancel the vertex that you added, you can delete it by pressing the Undo button

12. Continue digitizing for each of the buildings in the image.

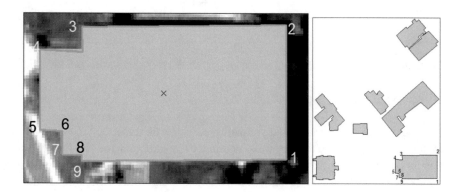

Result: the buildings in the images are digitized now

Update the Attribute Table of the Building

13. Open the attribute table of the Building layer
14. Highlight the first record and type 1, 2, 3,... under the "Id" field
15. In the Editor Toolbar click the Editor drop down arrow and click the Stop Editing

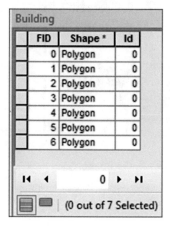

Note: The Building attribute table has 3-fields: FID, Shape*, and Id. A new field will be added to accommodate the name of the building.

15. Click Table Options/Add Field
16. Name: Name
17. Type: Text
18. Length 10
19. OK
20. In the TOC/R-click on the **Buildings** layer, click on **Edit Features** and then click on **Start Editing**.
21. Populate the field "Name" as seen below

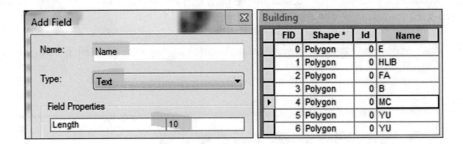

FID	Shape *	Id	Name
0	Polygon	0	E
1	Polygon	0	HLIB
2	Polygon	0	FA
3	Polygon	0	B
4	Polygon	0	MC
5	Polygon	0	YU
6	Polygon	0	YU

Create New Line Shapefile

You are now ready to begin digitizing the "Sidewalk"

20. Zoom In on the area south-east of the image, and you will see a sidewalk there.
21. In the Create Features window, click on the Sidewalk, and in the Construction Tools below, click on the line tool.
22. Using the aerial photo as a guide, digitize the new line by clicking the map each place you want to add a vertex
23. Continue clicking on the Sidewalk as you go along it to completely digitize it. Once you are done right click and click on Finish Sketch.

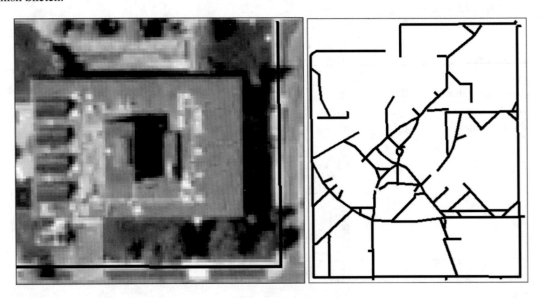

24. Continue digitizing for each of the Sidewalks in the image.

Create New Point Shapefile

Now we will digitize the tree layer. Begin by changing the symbol for the tree layer.

25. Double click on the symbol for the tree layer and then search "Tree" in the search bar. Then click on the desired tree symbol.

26. Now you will start digitizing by clicking on the Create Features window, and then clicking on "Trees" and then "Point" in the Construction Tools below.
27. Find a tree on the image background and click on it to digitize that point.
28. Repeat this process for the entire area of interest until you digitize all of the trees

29. With the Buildings, Sidewalk and Trees layers digitized, ArcMap will appear as follows for the area of interest. Note: Final digitizing of the Buildings, Sidewalk and Trees are in \Q6\Digitized folder

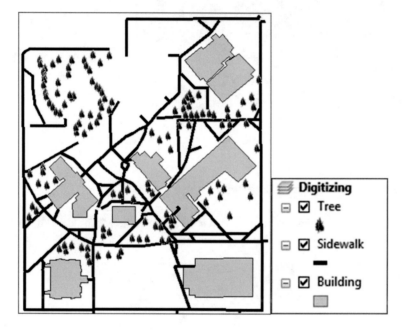

Update the Digitized Map

Overtime, some features will be removed, others will be modified, and new features will be added. To make these changes, sometimes you need to use an up-to-date image to update the features.

Scenario 7: The campus in 2010 modified, some building such as McCaskill Hall (MC) has been removed, Erlanson Hall (E) extended. And Yellowjacket Union (YU) student center modified and moved to another location. Two new buildings were added to the campus: Swenson Hall and the Greenhouse (GH). Your duty is to use an up-to-date image and perform all the changes.

GIS Approach
1. Insert Data Frame and call it **New Campus**
2. Click Add Data and browse to \\Data\Q6\Image highlight Campus_UTM15.tif and click Add
3. Click Add Data and browse to \\Data\Q6\Digitized and add Building.shp, Sidewalk.shp, and Tree.shp
4. R-click Building/and check Label Features
5. Change the symbol of the trees by choosing the Tree Symbol and make it size 10
6. Change the Sidewalk symbol by choosing "Expressway Ramp" and with a width of 2

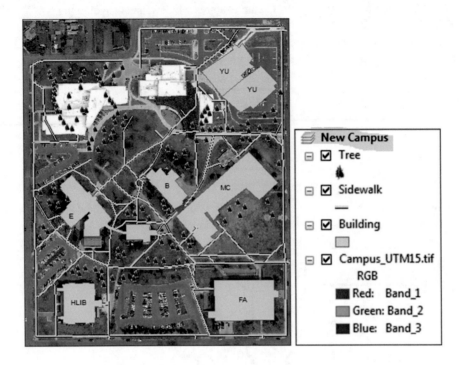

Result: The image with the buildings, sidewalks, and trees that was digitized in the previous steps is displayed.

Remove the Non-existing Buildings from the New Image

The first step is to remove the buildings that do not exist anymore such as the McCaskill Hall (MC) and the Yellowjacket Union (YU) student center that has been destroyed and constructed at another location.

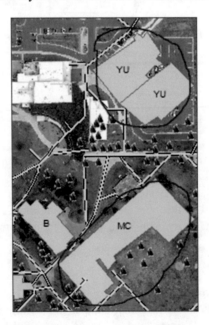

7. In TOC/R-click Building layer/Edit Features/Start Editing
8. In the Start Editing dialog box/highlight Building/click Continue
9. Zoom in around the YU building (you notice that the building does not exist in the new image, and a parking lot replaced it)

10. Click on the Edit Tool ▶ in the Editor Toolbar, then click on one of the YU building, click Shift and click on the second building (now both buildings are selected)

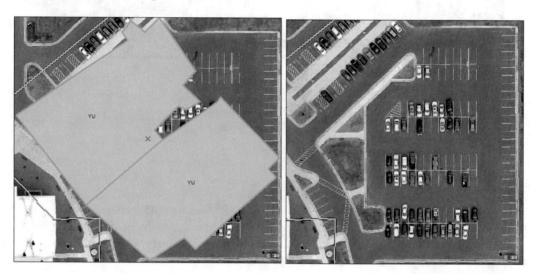

11. Click "Delete" on the Keyboard (Both of the YU buildings will disappear, and you will see that a parking lot replaced the buildings)
12. Zoom in around the MC building, you notice that the building does not exist in the new image and is replaced by Oexemann Greenhouse (GH)

13. Click on the Edit Tool ▶ in the Editor Toolbar, then click on MC building to select it and click "Delete" on the Keyboard (now the MC building is deleted)

Modify the E Building

The previous digitizing missed the southern part of Erlanson Hall (E) and your duty is to digitize it using the "Auto Complete Polygon" tool

14. Zoom in around the southern part of E Building.
15. Click on the **Create Features** window and then click on Buildings, and then Auto Complete Polygon under Construction Tools.
16. Click inside the E building first, then click on the upper right corner of the non-digitized building on the image, then on the lower right corner, then on the upper left corner and then Double-click inside the digitized E building to finish digitizing

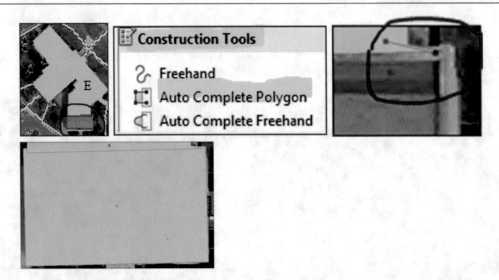

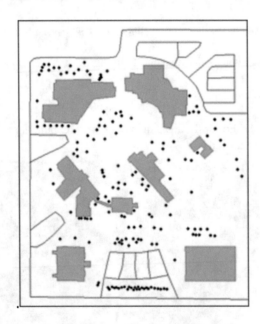

Next, you have to continue digitizing the Swenson Hall (SW) and Greenhouse (GH) buildings that have erected after the MC was building removed. You have to digitize them in order to have them added to the attribute table of the Building layer. After you finish digitizing the building, update the Tree and Sidewalk layers as seen in the image. Your final drawing should be like the one seen below

Coordinate Systems and Projections

Introduction

We are living on a spherical earth and the location of any feature can be determined by using a coordinate system.

The latitude and longitude is a 3-D coordinate system or spherical coordinate system that describes the location of features on the earth surface. Another type of coordinate system which can be used is called "Projected" or "Plane" coordinate.

There are various types of plane coordinates systems, which is expressed in x and y coordinate. The plane coordinates is the projection of the sphere from 3-D view into 2-D plane view. The latitude and longitude coordinate can be converted directly into different plane coordinates. The conversion from one coordinate to another is extremely important in any GIS work. The conversion from latitude longitude to any plane coordinate is called projection. Conversion from one plane coordinate to another plane coordinate is called re-projection. ArcGIS is an advanced software and can accommodate and handle any projections and coordinate conversions (Fig. 5.1).

Geographic Coordinate System (GCS)

The GCS is used to locate and measure the location of any feature on the earth surface in terms of latitude longitude and it's based on 3-D sphere. The earth is divided into two types of lines, **meridians** and **parallels**. Longitude is the line of meridians that run from north to south and measure the East-West locations. The prime meridian is running straight from the North Pole to South Pole and passing through Greenwich in England. The rest of the meridians are moving away from the prime meridian and are spaced farthest apart on the equator and converge to a single point at the North and South Poles.

Latitude are lines of parallels and run from East-West and measure location in North-South direction (Fig. 5.2). Parallels are equally spaced between the equator and the poles and always parallel to one another, so any two parallels are always the same distance apart all the way around the globe. Parallels and Meridians cross one another at right angles 90°.

The longitude ranges from 0° to 180° east and 0° to −180° west, while the latitude ranges from 0° to 90° north and 0° to −90° south. The East-North orientation is positive and the West-South orientation is negative.

The latitude longitude is used the same way the x, y coordinate is used in any plane coordinate. It is used as a reference grid to find location of features. The origin of the GCS is the point where the prime meridian intersects the equator.

Latitude and longitude can be measured either in degree minutes and second or decimal degree. One degree equals 60′ minutes and one minute equals 60″ seconds (see Chap. 4).

Electronic Supplementary Material: The online version of this chapter (https://doi.org/10.1007/978-3-319-61158-7_5) contains supplementary material, which is available to authorized users.

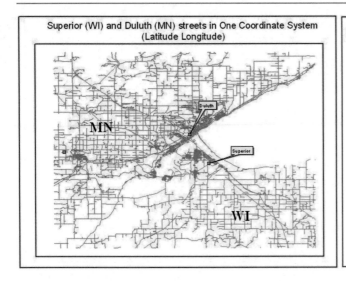

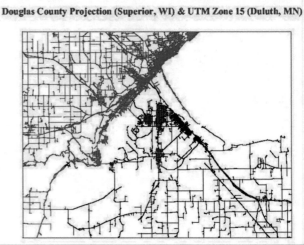

Fig. 5.1 *The left map* shows the streets of Superior (WI) and Duluth (MN) registered in latitude longitude. *The right map* shows the same file is registered in projected coordinate

Fig. 5.2 Line of latitude and longitude

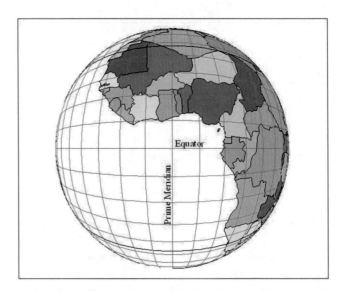

Map Projections

Map projection is a mathematical formula where the 3-D earth's transform into a 2-D or plane surface. A map projection is simply a systematic representation of a graticule of lines of latitude and longitude on a flat sheet of paper. The progress in computer technology and mapping mathematics makes the projection a challenging research.

Projection is used widely in cartography, land information system, remote sensing and GIS. The map projections generate different types of plane coordinates which are easier to use and work with than the spherical coordinate.

Projection and Distortion

The shape of the earth is so complicated and for the map projection, the earth is considered ellipsoid or sphere. The earth is rotated about its minor axis and has semimajor (a) and semiminor (b) axes (Fig. 5.3).

The shape, area, distance and direction of the features on the earth surface are correctly shown on a globe. The transformation from earth surface onto a flat plane surface involves distortion. Parallels, meridians, and the perpendicular intersection of parallels and meridian cannot be duplicated. The major alteration has to do with the angles which will affect the area, shape, distance and directions. There is no ideal projection that retains the major globe properties. This leads scientist to

Fig. 5.3 The shape of an ellipsoid

Pole

Semiminor axis

b

Equator

a

Semimajor axis

Pole

Fig. 5.4 Projection surfaces the cylinder and the cone

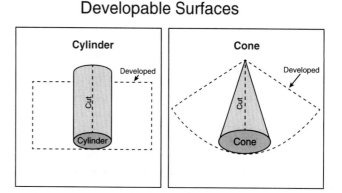

Developable Surfaces

generate hundreds of map projections in order to minimize the distortion and retain at least one of the earth's properties. Therefore, cartographer classified the map projection into major and minor properties. The major is the conformal and equivalent, the minor is the equidistant and azimuthal. Here are some examples about the major world projection.

1. **Conformal or orthomorphic projection** retains the angle and the shape of a small area. When this condition happens, the parallels and meridians will intersect at 90°.
2. **Equivalent or equal-area projection** will retain the correct relative size. Thus in such projection, the parallels and meridians will not intersect at 90°.
3. **Equidistant projection** conserves the distance between two points in a map. The scale must be the same as the principal scale on the reference globe from which the transformation was made.
4. **Azimuthal or true-direction projection** represents the part of the earth's directions correctly with straight line. But no projection can show direction so that the latitude and longitude are straight lines.

Many map projections include more than one property, which is extremely important in a small scale map.

Geographers use three physical surfaces for the construction of map projections. Developable surfaces (Fig. 5.4) such as the cylinder and cone and non-developable surface such as the plane. The cylinder and cone are not flat at the time the projection is created but can be flattened later by making an appropriate cut in the surface and unrolling it. Without stretching or tearing no distortion, when unrolled, distortion of the developable surface or of the pattern drawn on it occur.

Map projection may be produced by a light from three viewpoints: at the center, from infinity, and on the surface of the globe. For example, a light can be used to project the globe on a cylinder. The light also can be placed at any desired location and this gives rise to variations of the map projection.

Changing the location of the light source, modifies the characteristic of the resulting projection to the tangent or secant intersect on the cylinder and the cone. The first is the simple (tangent) case and the second is the secant case. The simple case results in one line and the secant case results in two lines. With the simple case, the projection surface (azimuthal plane, cylindrical or

Fig. 5.5 Use the cone to construct
a map projection

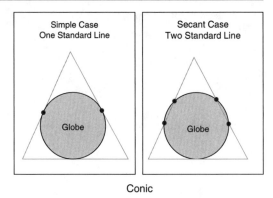

Fig. 5.6 Point and line of tangency

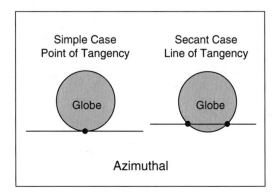

conic surface) touches the globe at one point or along one line. In secant case, the projection surface cuts through the globe to touch the surface at two lines (Fig. 5.5). The line of tangency is called a standard line in map projection. For cylindrical and conic projection the simple case has one standard line, whereas the secant case has two standard lines. If the standard line along the parallel is called standard parallel, and if it follows a meridian, it is called standard meridian. Along the standard line there are no distortions because there is a relationship one to one between the projection surface and the reference globe. The standard line is the identical to the reference globe, and away from the standard line a distortion occurs.

An azimuthal projection has only one point of tangency in the simple case and a line of tangency in the secant case. Therefore a plane may be tangent at any point on the globe (Fig. 5.6).

The orientation of the cylinder and the cone may be changed as desired and it can be normal, transverse and oblique. The normal orientation is when the cylinder can be placed so its tangent is along the equator and in the case of the cone its tangent along the parallel. Transverse Projection is when the cylinder or the cone is turned 90° from normal orientation. Oblique Projection is when the cylinder and cone lie between normal and transverse position.

Concept of the Datum

Datum is sets of parameters and ground control points defining local coordinate systems. Because the earth is not a perfect sphere, but is somewhat "egg-shaped," geodesists use spheroids and ellipsoids to model the 3-dimensional shape of the earth. Although the earth can be modeled by an egg-shaped solid, local variations still exist, due to differential thickness of the earth's crust, or differential gravitation due to density of the crustal materials.

A datum is created to account for these local variations in establishing a coordinate system. Figure 5.7 shows that the earth is irregular surface (thick black line). A generalized earth-centered coordinate system is called World Geodetic System 84 (WGS84). WGS84 provides a good overall mean solution for all places on the earth. However, for specific local measurements, WGS84 cannot account for local variations. Instead, a local Datum has been developed. For example, the local North American Datum of 1927 (NAD27, in dashed red line) more closely fits the earth's surface in the upper-left quadrant of the earth's cross-section. NAD27 only fits this quadrant, so to use it in another part of the earth will result in serious errors in measurement. For mapping North America, in order to obtain the most accurate locations and measurements, NAD27 is updated to NAD83 and NAD83 is adjusted in 1991. Any new map created in the U.S. will be based on NAD83.

Fig. 5.7 Datum (Image from ESRI)

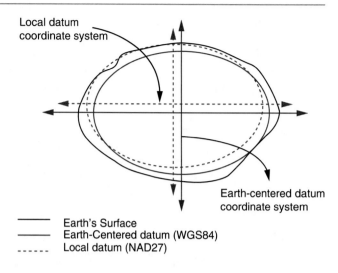

Local datum
coordinate system

Earth-centered datum
coordinate system

———— Earth's Surface
———— Earth-Centered datum (WGS84)
- - - - - - Local datum (NAD27)

Projection Parameters

A datum specifies the dimensions of a specific spheroid, a point of origin, an azimuth from the origin to a second point, and the spatial orientation of the spheroid relative to the earth. A GCS assigns unique coordinate values to locations on the surface of a spheroid. The system is usually based on latitude and longitude and is fully specified by a unit of measure (typically degrees), a prime meridian and a datum (e.g. NAD83). A Projected Coordinate System (PCS) is a combination of a map projection, projection parameters, and an underlying GCS that determines the set of X, Y coordinates assigned to a map.

When a map projection is used as a basis of coordinate system an origin should be established first. The point of origin is defined by the central parallel and the central meridian. The central parallel and central meridian in some literatures is called the latitude of origin and longitude of center respectively. For example Douglas County has the following projection parameters:

Douglas county of Wisconsin has the following parameters

Base Projection:	Transverse Mercator
Central Meridian:	-91.917
Central Parallel:	45.8833
False Easting:	194000
False Northing:	0.013 U.S. Survey feet
Scale Factor:	1.00004
Unit:	Foot US
Datum:	NAD83

Once map data is projected onto a planar surface, features must be referenced by a planar coordinate system. The latitude-longitude coordinate, which is based on angles measured on a sphere, is not valid for measurements on a plane. Therefore, a Cartesian coordinate system is used, where the origin (0, 0) is toward the lower left of the planar section. The true origin point (0, 0) may or may not be in the proximity of the map data you are using.

The following lessons will be performed in this chapter

1. Working with GCS
2. Projection on the Fly
3. Projection the GCS into UTM Zone 15
4. Georeferencing
5. Raster Projection
6. Datum Conflict

Working with GCS

The following two examples show how to calculate the distance in latitude-longitude coordinate.

Application 1: Distance calculation between two points on the map using the Pythagorean Theorem. The theorem is used to find the length of the hypotenuse of a right triangle, a calculation which affords many practical uses in various fields such as land surveying and navigation.

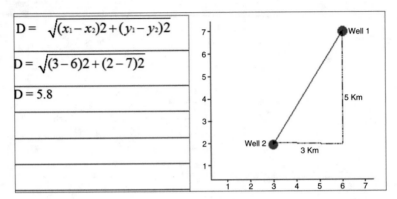

Application 2: Distance calculation between Superior and Eau Claire in Wisconsin using the latitude longitude. The location of the two cities are as follow:

1. Superior: −92.06 longitude and 46.70 latitude.
2. Eau Claire: −91.52 longitude and 44.80 latitude.

Formula of calculation: $\cos d = \sin a * \sin b + \cos a * \cos b * \cos c$

a = 46.7 latitude of Superior
b = 44.8 latitude of Eau Claire
c = 0.54 this is longitude of Superior − Longitude of Eau Claire
$\cos d = \sin 46.70 * \sin 44.80 + \cos 46.70 * \cos 44.80 * \cos 0.54$
$\cos d = 0.7277 * 0.7046 + 0.6858 * 0.7095 * 0.9999$
$\cos d = 0.999237$
d = 2.237 distance in degree
2.237 * 69.17 miles = 154.7 mile or (2.237 * 111.32 km = 249 km)

Projection on the Fly

On-the-fly projection is a powerful approach in GIS environment that allow users to combine data from any projection into a common projection for viewing and analysis. ArcGIS is armed with mathematics that can transform spherical coordinates to projected systems and vice versa on the fly without difficulties. The most important step to make the projection on the fly successful is to make sure that your file (if you are working with shapefile only) includes the *.prj file.

Scenario 1: You are an ecologist working to define certain species in Lake Menomonie. You have used the GPS and integrated the GPS file into ArcGIS. Your duty is to align the GPS data which is in latitude longitude with the image that has UTM coordinate system. In this exercise you are going to do two functions: (1) projection on the fly (2) project the GPS file permanently into UTM Zone 15N.

Define the Coordinate System and Datum

1. Start ArcCatalog
2. Click Customize menu points to ArcCatlog Options
3. Click Metadata tab / change Metadata Style into **FGDC CSDGM Metadata**

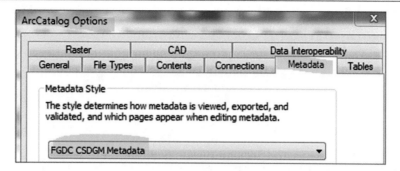

4. Click Apply and then OK
5. Hit F5 to refresh
6. Connect to your folder \\Ch05\Data\Q1 folder
7. Highlight in the Q1 folder "**lake.tif**"
8. Click the "**Description**" item in the right panel

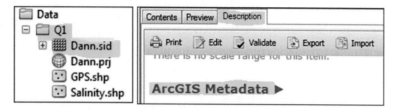

9. Scroll down and click "**ArcGIS Metadata**"
10. Click "**Spatial Reference**"

Result: **lake.tif** is an image projected into **UTM ᴢᴏɴᴇ 15** and its **DATUM** is **NAD83**.

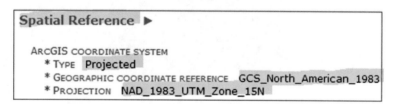

11. Highlight the **GPS.shp** in Q1 folder
12. Click Description Tab
13. Scroll down and click "**ArcGIS Metadata**"
14. Click "Spatial Reference"

Result: **GPS.shp** file is registered in **GCS** and its **DATUM** is **NAD83**.

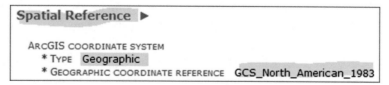

Note: **lake.tif** is projected and GPS.**shp** is not projected, but both represent the same area. ArcGIS can display both files on the fly.

Apply the Projection on the Fly

Projection on the fly means that you want to align two files from the same area together despite a difference in their coordinate system.

15. Launch ArcMap
16. R-Click the **Data Frame "Layers"**/Properties/Coordinate System
17. What is the coordinate of the Layers data frame? No coordinate system

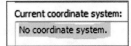

18. Integrate **lake.tif** into ArcMap from \\Data\Q1 folder
19. Now, what is the coordinate of the Layers data frame? NAD_1983_UTM Zone_15N

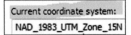

Comment: This means that the Data Frame inherited the coordinate system of the first layer integrated into ArcMap, which is the **lake.tif**.

20. Integrate the file **GPS.shp** into ArcMap from \\Data\Q1 folder
21. R-click the **GPS.shp** in the TOC
22. Point Zoom To Layer

Result: The two files are aligned with each other despite the **lake.tif** coordinate systems is different than GPS layer coordinate system.

Define the Projection

23. Click the **Add Data** ⊕▾ button and add **Salinity.shp** from \\Data\Q1 folder
24. Click Ok to accept that the **Salinity.shp** has missing spatial reference information

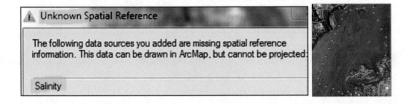

   Result: The **Salinity.shp** is not displayed in the extent of the **lake.tif** and **GPS.shp**, because the Salinity layer is missing spatial information. This means that the "**Salinity.shp**" is missing the file "**Salinity.prj**."

25. Click the Symbol of the **Salinity.shp**/chose Circle 2, size 12, cyan color, then click OK

26. Click the **Full Extent** button  on the *Tools* toolbar to see all the data in your map

How to Check That?

27. Open Catalog window
28. R-click Salinity.shp in \\Data\Q1 folder/Properties/click XY Coordinate System

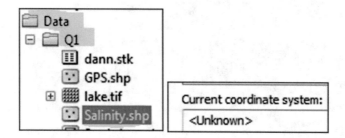

Result: Current coordinate system: <Unknown>. This means, the only information available is that the file has a latitude longitude but no spatial information (*.prj).

Open in window explorer the folder of \\Data\Q1, do you see the file "**Salinity.prj**"? The answer NO, therefore, you have to create it

29. Launch ArcToolbox by clicking ArcToolbox Window button  found on the Standard toolbar and duck it to the left of ArcMap (or click Geoprocessing menu/ArcToolbox)

Set the Environment

30. R-click an empty space in ArcToolbox
31. Environment/Workspace/
32. Current Workspace: \\Ch05\Data\Q1
33. Scratch Workspace: \\Ch05\Results
34. OK
35. In ArcToolbox, click Data management Tools/Projections and Transformations
36. D-click Define Projection
37. Browse to Salinity.shp in in \\Data\Q1 folder
38. Click the spatial reference icon to choose the coordinate system/select/D-click Geographic Coordinate Systems/North America
39. Select NAD 1983/OK/OK
40. R-click Salinty.shp point to Zoom To Layer

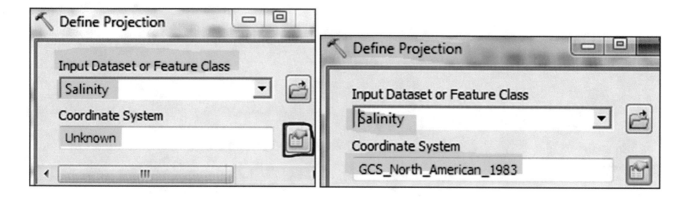

What is the result? Projection on the fly and the following happened:

a. The Salinity.shp is aligned with the **lake.tif** and **GPS.shp**
b. **Salinity.prj** a new file is created and added to the \\Data\Q1 folder
c. The Salinity.shp points are located inside Lake Menomonie

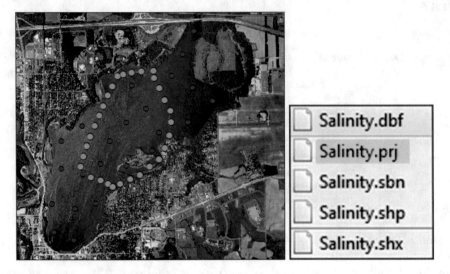

Projection the GCS into UTM Zone 15

The **lake.tif** and **GPS.shp** have different coordinate system. **lake.tif** has a UTM projection zone 15N and **GPS.shp** has a geographic coordinate system (latitude and longitude). You are interested in projecting the **GPS.shp** from latitude-longitude coordinate system onto UTM zone 15 N.

1. Insert new Data Frame and call it "**Projection**"
2. Integrate the **lake.tif** and **GPS.shp** into ArcMap from \\Data\Q2 folder
3. Activate ArcToolbox if it is not active from the previous exercise
4. In ArcToolbox, click Data management Tools/Projections and Transformations
5. D-Click Project
6. Drag **GPS.shp** from the Table of Content and place it into the Input Dataset in the Project Dialog Box
7. Output Dataset **GPS_UTM.shp** and save it in \\Result folder
8. Output Coordinate System: Click the button for the coordinate system to open the Spatial Reference Properties Dialog box
9. D-Click 'Projected Coordinate Systems'/UTM/NAD 1983
10. Click NAD 1983 UTM Zone 15N
11. Click OK/OK

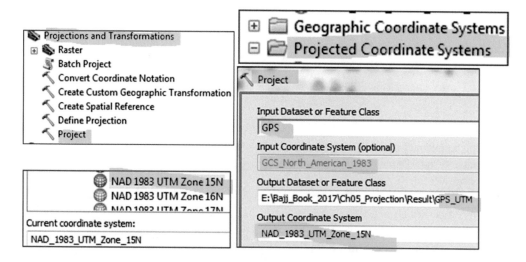

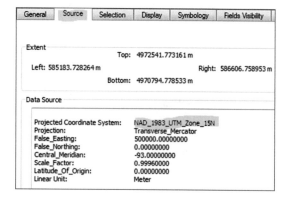

12. **GPS_UTM.shp** added from \\Result folder into the TOC
13. R-click **GPS_UTM.shp** in the TOC/Properties/Source

Result: the **GPS_UTM.shp** is now registered in **NAD_1983_UTM_Zone_15N**.

Georeferencing

Raster data is commonly obtained by scanning maps or collecting aerial photographs and satellite images. Scanned maps and some downloadable images don't usually contain information as to where the area represented on the map fits on the surface of the earth; the locational information delivered with aerial photos and satellite imagery is often inadequate to perform analysis or display in proper alignment with other data. Thus, in order to use these types of raster data in conjunction with your other spatial data, you often need to align or georeference it to a map coordinate system.

When you georeference your raster, you define how the data is situated in map coordinates. This process includes assigning a coordinate system that associates the data with a specific location on the earth. georeferencing raster data allows it to be viewed, queried, and analyzed with other geographic data.

Scenario 2: You are an ecologist and you are interested in the wetland in eastern Wisconsin. One of the critical issues in your research is to determine the hydraulic relationship between the groundwater of the Silurian dolomite aquifer and the wetland. In order to conduct the research, you have to collect any type of data to support your research. You have found on the internet an image representing Silurian aquifer in east Wisconsin. The image has a false coordinate and the rest of your digital data is registered to a geographic coordinate (Latitude-Longitude). Your first target is to georeference the image using the State of Wisconsin file.

You can obtain the state of Wisconsin file using the **ESRI databas**e that come with ArcGIS software. In this exercise you will be given the file.

GIS Approach

1. Insert Data Frame and call it "**Georeference**"
2. Add **State48**.**shp** from the folder \\Ch05\Data\Q3

Set Environment

3. Click Geoprocessing menu and point to Environment and click Workspace
    1. Current Workspace: \\Ch05\Data\Q3
    2. Scratch Workspace: \\Result
4. Click OK
5. R-click **State48**.**shp**/Properties/Source

   Result: the _coordinate_ & _datum_ of the **State48**.**shp** is _GCS_North_American_1983_.

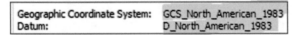

| Geographic Coordinate System: | GCS_North_American_1983 |
| Datum: | D_North_American_1983 |

6. After viewing the coordinate close the **Layer Properties** dialog box
7. R-click States48 layer in the TOC/Label Features

   Result: The states is now labeled by the name field.

8. Use the **Select Features** tool  on the *Tools* toolbar
9. Click state of Wisconsin to select it
10. R-click **State48**.**shp** point to **Data** select **Export Data**
11. Save it as **WI**.**shp** in \\Result folder

12. **OK/Yes** to add **WI**.**shp** to the **TOC**
13. R-click **State48**.**shp** in the **TOC** point to **Remove**
14. R-click **WI** layer in the **TOC**/Zoom To Layer
15. Click Add Data and integrate the raster "**WI_Aquifer**.**jpg**" from \\Data\Q3 folder
16. Click Ok on the note saying that

   The following data sources you added are missing spatial reference information. This data can be drawn in ArcMap, but cannot be projected

Result: The "**WI_Aquifer.jpg**" display's far from the **WI** layer, because the image has false coordinate system.

17. Click Customize/Toolbars and check Georeferencing

The Georeferencing tool displays and the **WI_Aquifer.jpg** shows in the Layer drop down.

Note: if you have more than one raster, click the **Layer** drop-down arrow and click the raster layer you want to Georeference.

18. Click Georeferencing in the Georeferencing tool and click **Fit To Display**. This will display the raster in the same area as the **WI** layers.
19. Click on the **WI** layer symbol color and choose "Hollow" and red to outline color

Note: This displays the raster dataset in the same area as the target layers. You can also use the **Shift**

and **Rotate** tools to move the raster dataset as needed. To see all the datasets, adjust their order in the table of contents.

Results: You see now both the image and vector together.

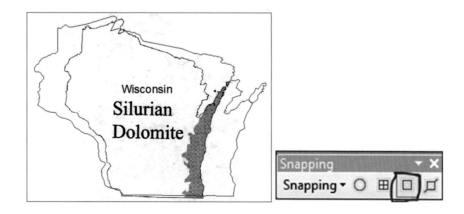

20. Click Editor toolbar/Snapping/Snapping toolbar
21. Activate Snapping tool in Snapping toolbar (this snap to the vertex of WI layer)

22. In the Georeferencing tool click the Add Control Points tool 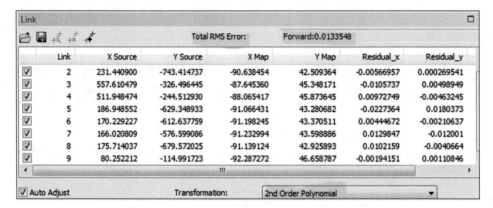 to add control points.

23. To add a link, click the mouse pointer over a known location on the raster, then over a known location on the target data.
24. Add enough links for the transformation order. You need a minimum of three links for a first-order transformation, six links for a second order, and 10 links for a third order

   Note: It is recommended to add at least 8 points, from south, east, north and west of the image. Try to use a known location such as corner or intersection. Try also each time you add a control point to zoom in, in order to achieve more accuracy (9 points selected).

25. Click View Link Table in Georeferencing toolbar
26. Change the transformation into second Order Polynomial
27. Make sure Auto Adjust is checked

	Link	X Source	Y Source	X Map	Y Map	Residual_x	Residual_y
☑	2	231.440900	-743.414737	-90.638454	42.509364	-0.00566957	0.000269541
☑	3	557.610479	-326.496445	-87.645360	45.348171	-0.0105737	0.00498949
☑	4	511.948474	-244.512930	-88.065417	45.873645	0.00972749	-0.00463245
☑	5	186.948552	-629.348933	-91.066431	43.280682	-0.0227364	0.0180373
☑	6	170.229227	-612.637759	-91.198245	43.370511	0.00444672	-0.00210637
☑	7	166.020809	-576.599086	-91.232994	43.598886	0.0129847	-0.012001
☑	8	175.714037	-679.572025	-91.139124	42.925893	0.0102159	-0.0040664
☑	9	80.252212	-114.991723	-92.287272	46.658787	-0.00194151	0.00110846

Total RMS Error: Forward:0.0133548

☑ Auto Adjust   Transformation: 2nd Order Polynomial

Root Mean Square (RMS): The error report includes two different error calculations: a point-by-point error and a root mean square (RMS) error. The point-by-point error represents the distance deviation between the transformation of each input control point and the corresponding point in map coordinates. The RMS error is an average of those deviations.

28. If you're satisfied with the registration, you can stop entering links.
29. If you're not satisfied with the registration, you can delete the control points or some of the points
30. To do that click Georeferencing Menu/Delete Control points
31. Repeat the previous steps to start Georeferencing once more time
32. Once you finished and you are satisfied with the RMS
33. The Total RMS Error is **0.013** (this is acceptable)

34. Click Georeference menu **Rectify** (This creates a new file with the same name as the raster but with an .aux file extension).
35. Browse and save the image in the \\Result folder

36. Call the image **Rec_WI_Aquifer.tif** and click Save
37. Click Georeferencing toolbar and click **Update Georeferencing** to save the transformation information with the raster.

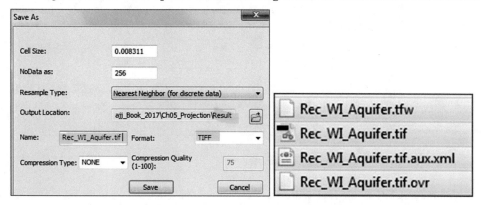

38. R-click **WI_Aquifer.tif** in the **TOC** and **Remove**
39. Click Add Data and add **Rec_WI_Aquifer.tif** from \\Result folder

Result: The **Rec_WI_Aquifer.tif** aligned with WI layer.

Exercise: it is recommended to perform the georeferencing using a projected layer. Therefore, you have been asked to project **WI.shp** into WI Transverse Mercator (**NAD 1983 Wisconsin TM (Meters)**) and call it WI_TM, and then use the projected layer to Georeference the **WI_Aquifer.jpg** once again.

Raster Projection

This exercise allow you to project the raster "**Rec_WI_Aquifer.tif**" that you have georeferenced in the previous section.

Scenario 3: You have decided to use the raster **Rec_WI_Aquifer.tif** to manage the groundwater in eastern Wisconsin. In order to do this you have decided to project the file into Wisconsin Transverse Mercator (WTM).

1. Insert Data Frame and call it "Raster Projection"
2. Click Add Data and add **Rec_WI_Aquifer.tif** from \\Data\Q4 folder

Result: The **Rec_WI_Aquifer.tif** displays with black background.

3. R-click **Rec_WI_Aquifer.tif** in the TOC/Properties/click Symbology tab
4. Highlight Stretched (under Show) and check Display Background Value

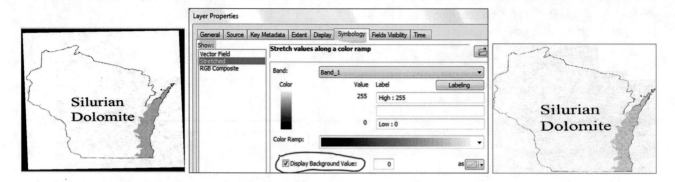

5. Click Source tab and scroll down to the **Special Reference**

    Result: the Coordinate is GCS (latitude-longitude) and the Datum is NAD83.

Spatial Reference	
XY Coordinate System	GCS_North_American_1983

6. Click OK to exit the Layer Properties dialog box

7. Launch ArcToolbox by clicking ArcToolbox Window button 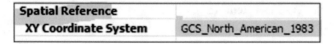 found on the Standard toolbar and tuck it to the left of ArcMap (if it is not already there)

Set the Environment

1. R-click an empty space in ArcToolbox
2. Environment/Workspace/
3. Current Workspace: \\Ch05\Data\Q4
4. Scratch Workspace: \\Result
5. OK

Project the Rectified Raster

This tool will transform the raster dataset from one coordinate system to another. In this exercise, it will transform from latitude–longitude into Wisconsin Transverse Mercator.

6. In ArcToolbox, click Data management Tools/Projections and Transformations
7. Under **Projection and Transformation**
8. Open the **Raster** and double click **Project Raster**

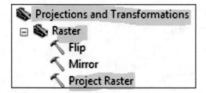

9. The Project Raster dialog box will display
10. Drag the **Rec_WI_Aquifer.tif** from TOC and place it in the Input Raster
11. Output Raster Dataset browse to \\Result and call the raster "**Aquif_WTM.tif**"

12. Click the Spatial Properties  button
13. Open Projected Coordinate Systems/open State Systems/
14. Scroll down to the end and select "**NAD 1983 Wisconsin TM (Meters)**"

🌐 NAD 1983 Wisconsin TM (Meters)

15. Click OK/OK

Result: The "**Aquif_WTM.tif**" is displayed in the TOC and it is now projected and has the customized Transverse Mercator of Wisconsin.

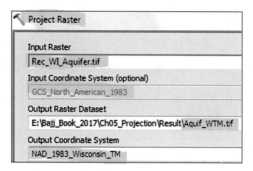

16. R-click **Aquif_WTM.tif** in the TOC/Properties/click Source tab
17. Scroll down to the **Special Reference**

Result: the Coordinate is Wisconsin Transverse Mercator (Wisconsin_TM) and the Datum is NAD83.

Spatial Reference	
XY Coordinate System	NAD_1983_Wisconsin_TM

Datum Conflict

Some countries have more than one coordinate system and datum. Often the times when you re-project the features from one coordinate to another, the conversion is not always accurate.

Scenario 4: You are a new employee at the Water Authority of Jordan and you have given an Excel file (**Well.xls**) that include the well numbers and their coordinates in latitude-longitude from Dhuleil Area. You have been asked to do the following:

1. Integrate the Wells (**Well.xls** file) into ArcGIS and assign to it the Palestinian Datum and save it as a shapefile and call **Well_PD.shp**
2. Project the **Well_PD.shp** into "**Palestine_Belt**" and call it "**Well_PTM.shp**"
3. Integrate the Well.xls file into ArcGIS again and assign to it the Jordanian Datum and save it as a shapefile and call it **Well_JD.shp**
4. Project the **Well_JD.shp** into "**JTM**" and call it "**Well_JTM.shp**"
5. Re-project the **Well_JTM** into **Well_PTM.shp**

GIS Approach

1. Insert Data Frame and call it "**Datum Conflict**"
2. Add **Dhuleil.shp** into ArcMap from \\Data\Q5 folder

3. R-click **Dhuleil.shp**/Properties/Source
4. The coordinate of the **Dhuleil.shp** is a Customize UTM Coordinate

Projected Coordinate System:	Palestine_1923_Palestine_Belt
Projection:	Transverse_Mercator

Now, you should add the Excel file (Well.xls) that has two fields of the coordinate system in latitude and longitude.

No	Long	Lat
1	36.3504	32.1415
2	36.2095	32.1674
3	36.3024	32.1543
4	36.3118	32.1736
5	36.3119	32.1499
6	36.3039	32.1488
7	36.2923	32.1511
8	36.3951	32.129
9	36.3443	32.1354
10	36.3212	32.1367

5. Click Add Data and browse to the \\Data\Q5 folder
6. D-click Well.xls/select Data$ and click Add
7. The Data$ added to the TOC
8. R-click Data$/Display XY Data
   a. X Field: Long
   b. Y Field: Lat
   c. Click Edit/Geographic Coordinate Systems/Asia/Palestine 1923
   You are selected the **Palestine 1923** datum that associated with the latitude-longitude
9. OK/OK/Yes to create Object-ID

Result: The wells will display inside the ArcMap in the Data View.

10. R-click Data$Events/Data Export call it **Well_PD.shp** in the \\Result folder

11. Click Yes to add the **Well_PD.shp** into the TOC
12. Remove **Data$Events** and **Data$** from the TOC

Project the Well_PD from Latitude-Longitude into Palestinian Projection

13. Insert Data Frame and rename it Palestine Projection
14. Copy **Dhuleil.shp** and **Well_PD.shp** into the Palestine Projection Data Frame
15. Launch ArcToolbox/Set the Environment
    a. Current Workspace \\Data
    b. Scratch Workspace \\Result
16. OK
17. ArcToolbox/Data Management Tools/Projections and Transformations\
18. D-click Project Tool
19. Input Dataset Or Feature Class: Well_PD
20. Output Dataset Or Feature Class: \\Result\Well_PTM
21. Output Coordinate System: Projected Coordinate Systems\National Grids\Asia\Palestine 1923 Palestine Belt
22. OK/OK

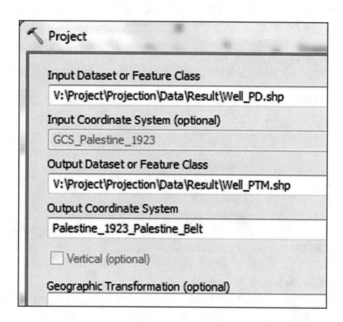

Note: The Palestine Belt Projection is based on the Transverse Mercator and it parameters are:

False_Easting:	170252
False_Northing:	1126868
Central_Meridian:	35.2121
Scale_Factor:	1
Latitude_Of_Origin:	31.7341
Linear Unit:	Meter

Project the Data into Jordanian Projection

23. Insert Data Frame and rename it Jordanian Projection
24. Add **Dhuleil.shp** and then browse to the folder \\Data\Well.xls\Data$
25. R-click Data$
    a. X Field: Long
    b. Y Field: Lat

   c. Click Edit/Geographic Coordinate Systems/Asia/Jordan
26. OK/OK/OK
27. R-click Data$Events/Data Export \\Result\Well_JD.shp
28. ArcToolbox/Data Management Tools/Projections and Transformations\
29. D-click Project Tool
30. Input Dataset Or Feature Class: **Well_JD**
31. Output Dataset Or Feature Class: \\Result**Well_JTM**
32. Output Coordinate System: Projected Coordinate Systems\National Grids\Asia\Jordan JTM
33. OK/OK

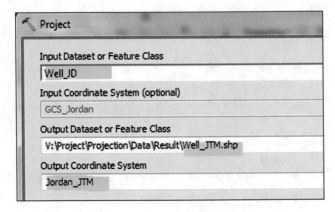

Note: The Jordanian Projection is based on the Transverse Mercator and its parameters are:

False_Easting:	500000
False_Northing:	-3000000
Central_Meridian:	37
Scale_Factor:	0.9998
Latitude_Of_Origin:	0
Linear Unit:	Meter

Project the wells from Jordanian Projection into Palestinian Projection

34. Insert Data Frame and rename it Jordanian_Palestinian
35. Copy **Well_PTM.shp**, **Dhuleil.shp** and **Well_JTM.shp** into the Jordanian_Palestine Data Frame

   Result: Projection on the fly took place.

36. D-click Project Tool
37. Input Dataset Or Feature Class: **Well_JTM**
38. Output Dataset Or Feature Class: \\Result\Well_JTM_PTM.shp
39. Output Coordinate System: import from the file \\Result\Well_PTM.shp
40. Add\OK\OK
41. Geographic Transformation:
   a. Jordan_To_WGS_1984
   b. Palestine_1923_To_WGS_1984_1
42. OK

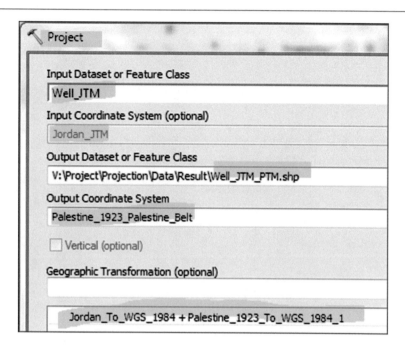

Comment: The **Well_JTM_PTM** is added to the TOC. Nevertheless, ArcGIS software can't proceed with the projection unless the following happened: The software needs to convert first the Jordanian datum into WGS84, then the Palestine datum to WGS84.

   Result: the result of the projection shows that the projected **Well_JTM_PTM.shp** didn't align correctly with the location of the **Well_PTM.shp**. The discrepancies between the same well from the two files around 177 m. This could be to the accuracy of the Palestine or Jordanian datum.

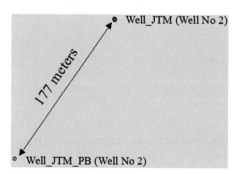

How to work around the problem

The only way to work around this problem is to perform the transformation through a geodatabase approach, which will be discussed in the Geodatabase chapter.

ArcScan

6

Introduction

ArcScan is another way to get digital data that can be used in GIS. The ArcScan can generate digital data in vector format from a scanned image. The ArcScan extension in ArcGIS is designed for vectorization, which means converting raster data to vector data. The process of vectorization (tracing) can be performed either manually or automatically.

Manual vectorization requires that the user trace the raster that is subject to conversion. The automated procedure requires that you generate features for the entire raster based on settings that you specify. ArcScan provides an efficient way to streamline this integration when compared to traditional techniques, such as digitizing.

Since ArcScan is designed to work with editing, users must start an edit session to activate the ArcScan toolbar. This means that all the editing tools and commands can be used in conjunction with the ArcScan tools and commands. ArcScan uses editing settings, such as the snapping environment and target templates and layers.

ArcScan is a powerful tool and it is available as an add-on component of the ArcGIS Desktop suite. Licensed as a separate extension, ArcScan works within the ArcMap environment and relies on its own user interface, which supports the tools and commands used for the vectorization process.

Important Note 1: ArcScan can vectorize any raster format supported by ArcGIS as long as it is represented as a bi-level image. If you have more than one band image you have to integrate one band only.

Important Note 2: if you want to make your multiband only 1 band, you have to do the following:

- Add to ArcMap the 1 band only from the multiband image
- Export the 1 band (Data/Export) and call it a new name
- Launch ArcToolbox/Spatial Analyst/Reclass/Reclassify
- Reclassify the one band image into two classes.

Scenario: You are a hydrogeologist and you are interested in converting the georeferenced image from Azraq Basin into a vector digital data file format. This conversion will allow you to view the stream system as a vector layer and then use it in you analysis in your project.

To use the ArcScan in this lesson you need to use the rectified image "**Rec_Stream.tif**" from your folder. The image represents stream system in the central part of Azraq Jordan. The main duty is to perform an automatic conversion with high precision from raster into vector.

Electronic Supplementary Material: The online version of this chapter (https://doi.org/10.1007/978-3-319-61158-7_6) contains supplementary material, which is available to authorized users.

© Springer International Publishing AG 2018
W. Bajjali, *ArcGIS for Environmental and Water Issues*, Springer Textbooks in Earth Sciences, Geography and Environment,
https://doi.org/10.1007/978-3-319-61158-7_6

Data Requirement

1. **Rec_Stream.tif**, the image exist in \\Data\Image folder
2. Create a line file and call it **Stream.shp** and save it in \\Data\Vector folder
3. ArcScan toolbar to vectorize the **Rec_Stream.tif**

Data Integration

1. Start ArcMap
2. Click Add Data/connect to Ch06
3. Browse to \\Data\Image folder and select **Rec_Stream.tif**
4. Click Add

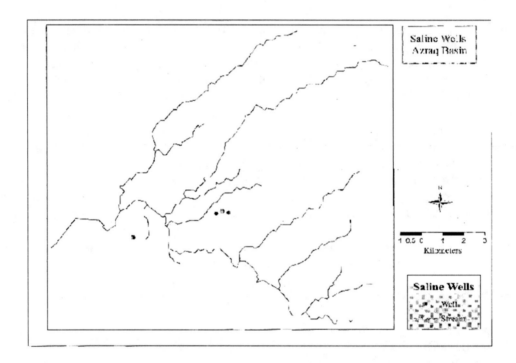

Result: The **Rec_Stream.tif** (consists of 1 band) is added to the TOC.

5. In the TOC/R-click **Rec_Stream.tif**/Properties/Symbology tab
6. Under **Show**: highlight the Unique Values (you see the image consists of 255 classes) Click OK to exit the dialog box

Note: Your duty is to make the image to have only two classes (0, 1).

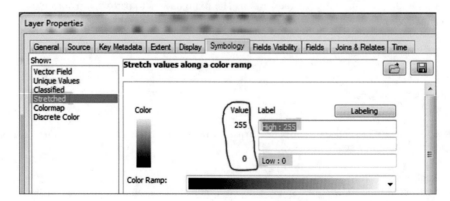

Make a Bi-level Image with 2-Classes by Reclassification

This step aims to reclassify the **Rec_Stream.tif** image into two classes permanently

7. Click on the ArcToolbox ![ArcToolbox icon] button on the Standard Toolbar and place it in the left panel
8. R-click an empty place in the ArcToolbox/click Environment/click Workspace
9. Fill it as below
10. Current Workspace: \\Ch06\Data\Image
11. Scratch Workspace: \\Result
12. Click OK

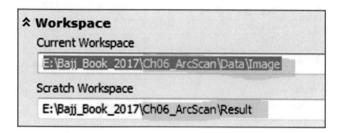

Use Reclassify Tool to Make the Image to Have Two Classes (0, 1)

Note make sure the Spatial Analyst Extension is checked

13. ArcToolbox/Spatial AnalystTools/Reclass/D-click Reclassify Tool

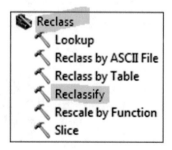

14. Fill the Reclassify dialog box as below
15. Input raster **Rec_Stream.tif**
16. Reclass field **Value**
17. Output raster \\Result\Reclass_Str.tif
18. Click Classify

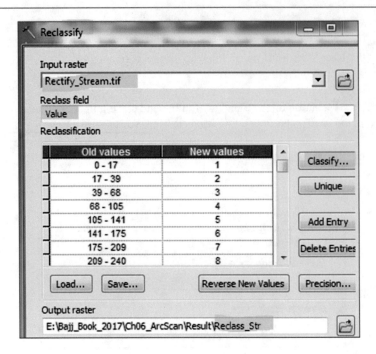

19. Method: Manual
20. Classes: 2
21. Click OK/OK

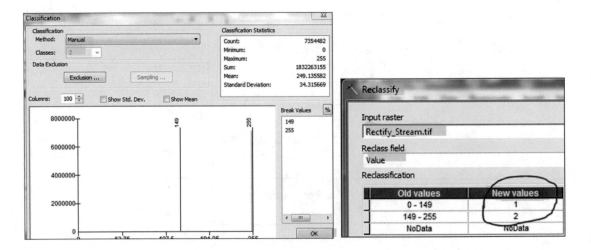

Result: The **Reclass_Str.tif** is added into TOC and now it is a bi-level raster with 2-classes.

Reclassification of the Image

Now, you are going to classify the **Reclass_Str.tif** into a two classes with black and white color

22. R-click **Reclass_Str.tif**/Zoom to Layer
23. R-click the **Reclass_Str.tif**/Properties/Symbology tab/select "Classified" under the Show window.

24. Under "Label" change the first value to **0** and second value to **1** and make sure the symbol of the 0 label is black and the symbol of 1 label is white.

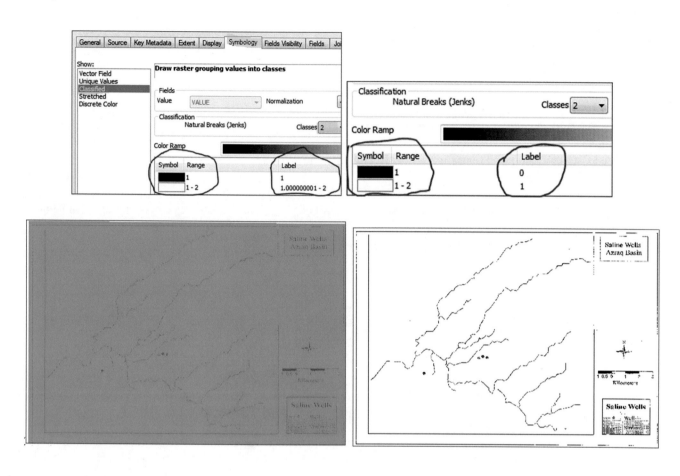

Result: The **Reclass_Str.tif** image display in 2 classes 0 (black) is ink and 1 (white) no ink.

Prepare the Image for Vectorization

Prior to vectorization you have to **Activate ArcScan**

26. Click **Customize menu/**point to **Extensions/**check **ArcScan**
27. Click **Customize menu/**point to **Toolbars/**click **ArcScan**

Result: The ArcScan toolbar display with the "**Reclass_Str**" image in the ArcScan window.

Create a Blank Shapefile

This step is to create a line shapefile to be used during vectorization

28. In Catalog window/R-click \\Result folder/New/Shapefile
29. Name: **Stream.shp**
30. Feature type: **Polyline**
31. Click Edit

32. Click Add Coordinate System 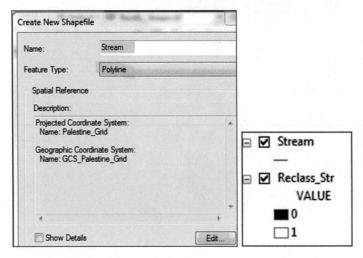 dropdown arrow
33. Click Import
34. Browse to \\Ch06\Data\Image folder and select "**Rectify_Stream**.tif"/Add
35. OK/OK

Result: The "**Stream.shp**" is now registered in Palestine_Grid, which is customized Transverse Mercator and will be integrated into the TOC.

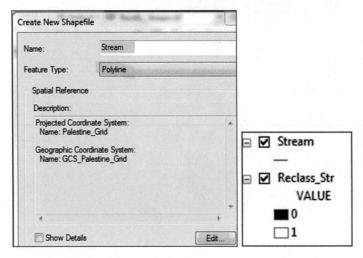

Raster Cleanup

This step is necessary to clean the unwanted information in the **Reclass_Str** in order to perform the vectorization. Raster cleanup will delete any objects or text in the image that required to be removed. The current **Reclass_Str** image includes the following that need to be deleted

- Border of the study area
- 4-wells
- Title in a border
- Scale bar
- Legend
- North arrow

This step is essential to the vectorization process especially if you want to run the automatic generating feature. The final step of the image cleaning is to keep only the stream network. The stream then will be converted into vector shapefile.

To start raster cleanup session, you must meet the following two requirements for using ArcScan.

a. Having a vector in editing mode and a raster layer
b. The raster layer should be symbolized in two values (classes).

Once the cleanup session is started, the **Raster Cleanup** menu commands will be available.

Remove the Undesired Objects in the Image

First Approach

36. In the TOC/R-click Stream/Edit Features/Start Editing
37. Once you start editing the **Reclass_Str.tif** raster will be activated in the **ArcScan** window

38. Click the **Raster Cleanup** menu on the ArcScan toolbar/click **Start Cleanup**
39. Click the **Raster Cleanup** menu on the ArcScan toolbar and point to **Raster Painting** toolbar

Result: The **Raster Painting** toolbar will be displayed.

Note: The **Raster Painting** toolbar supports **9 tools** designed for drawing and erasing raster cells. In this exercise, we are going to use only 2-tools the "**Erase tool**" and the "**Magic Erase tool**" . The Erase tool can delete cells and you can erase small areas in the raster by clicking with the tool and larger areas by clicking and dragging the erase cursor over a series of raster cells. The **Magic Erase tool** allows you to erase connected cells. You can drag a box around a series of connected cells to erase them. All connected cells that are completely within the box are removed.

Using Erase Tool

40. Click an Erase tool on the Raster Painting toolbar and then click on the **Erase size** tool and select the **largest size**
41. Zoom in to the upper right corner of the outer border
42. Place the Eraser tool on the upper point corner of the outer border and start erasing the border
43. Once you finished erasing the border, erase the border of Saline Well text

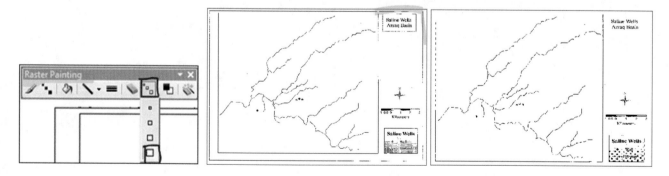

Result: The outer border and the border of the text are now erased.

Using Magic Erase Tool

44. Click the Magic Erase tool ![icon] .
45. Click the Magic Erase tool and drag a rectangle around the text of Saline Well Azraq Basin to erase it.
46. Use the Magic Erase tool again to drag a rectangle around north arrow, scale, legend, and the 4 wells to erase them

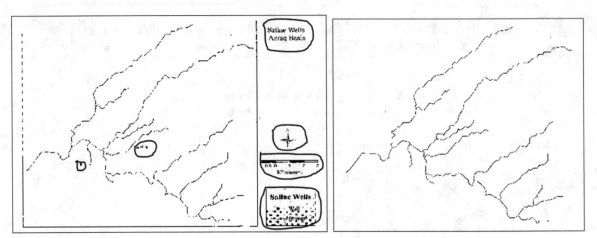

Result: All the obsjects are now erased.

47. Click the **Raster Cleanup** menu on the ArcScan toolbar and point to **Stop Cleanup** toolbar
48. Click Yes to Save your Raster Cleanup edits

Second Approach

49. Click the **Raster Cleanup** menu on the ArcScan toolbar and point to **Start Cleanup** toolbar

50. Click Select Connected Cells button ![icon] in ArcScan Toolbar
51. Click on the catchment border, and the border of the catchment area will be selected and turn to cyan color
52. Click Raster Cleanup menu on the ArcScan toolbar/Erase Selected Cells

Result: All the border will be removed.

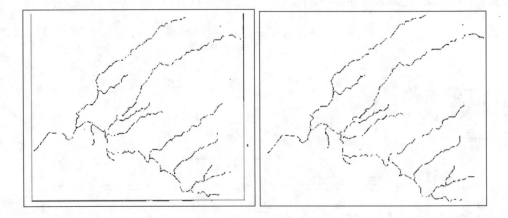

53. Select Raster Cleanup/Save/
54. Select Raster Cleanup again/Stop Cleanup

Result: The raster now has been cleaned and it is ready for vectorization.

Vectorization

This step allows you to convert the raster into a vector. There are two ways to vectorize the image: Interactive and Automatic Vectorization.

Interactive Vectorization (Vectorization Trace)

Interactive vectorization involves the manual creation of features assisted by the ability to snap to raster cells and utilization of the raster tracing and shape recognition tools. It is used when a user wants total control of the vectorization process or needs to vectorize a small area of the raster image. This approach allows a higher level of flexibility since you can use the ArcScan trace tools or the Create Features window construction tools and Editor toolbar sketch construction methods to generate features. You are going to trace one tributary (as seen below) to learn how to use this technique.

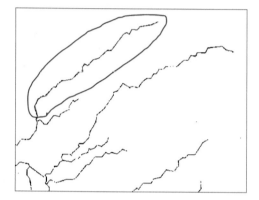

Vectorization Trace

Before starting the **vectorization trace** make sure you are still in editing mode.

1. Choose **Vectorization Setting** from the **Vectorization menu** at **ArcScan Tool**

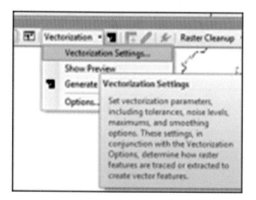

2. The Vectorization Setting dialog box display should be changed in the following:
   a. Intersection solution: None
   b. Maximum Line Width: 10
   c. Keep the rest without modification

Then Click Apply\Close

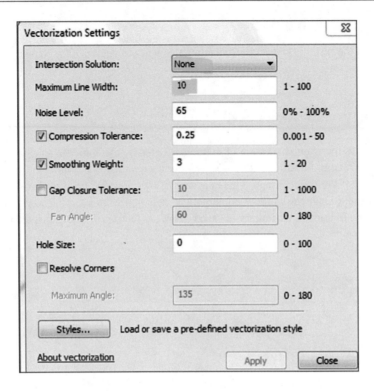

Raster Snapping

3. Click the Raster Snapping Options 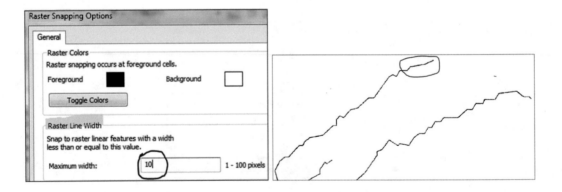 icon on the ArcScan toolbar to open the Raster Snapping dialog box. Under Raster Snapping Options
   (a) Set the maximum line width value to 10.
   (b) Accept the rest as default
   (c) Click OK
4. Zoom around the top of the first stream

5. To create lines, click the **Stream feature** template in the *Create Features* window

Vectorization Trace

6. Click the **Vectorization Trace** tool icon on the *ArcScan* toolbar
7. Hold the 'S' letter on the keyboard and click at the beginning of the stream, and continue clicking along the stream
8. Zoom in or use the Pan icon to move to the next stream segment
9. When you have reached the end of the intersection with other stream of the raster double click or press F2 to finish the sketch (**It will appear Cyan**)

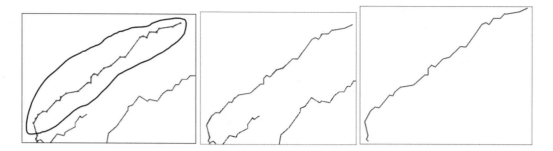

Note 1: the 'S' letter means **suspend the automatic tracing** and it is used when you encounter a break in the line of the image.

Note 2: in order to continue digitizing the tributary you have to use the Snapping toolbar. The Snapping toolbar is armed with four snapping types: points, end, vertex, and edges. Snapping icons allows you to create features that connect to each other with minimum errors.

Snapping Toolbar

10. On the Editor toolbar in ArcMap/click drop-down editor and point to Snapping and check *Snapping* toolbar.

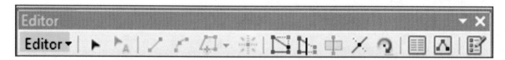

11. On the *Snapping* toolbar, click the **Snapping** menu and confirm that **Use Snapping** is checked.

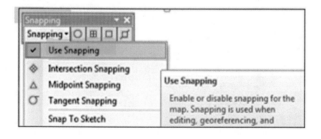

12. Click the Vertex Snapping ⬜ button

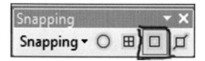

13. When you finish digitizing the whole the designated tributary/in TOC/R-click Stream/Open Attribute Table

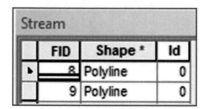

Result: You see the segments of the stream that you digitized. Because you are going to use the automatic digitizing, you are going to delete the stream.

14. Highlight the record in the attribute table and click the Delete Selected 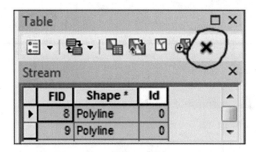 button

15. R-click "Stream"/click Editor/**Stop Editing**

Note: You can shift to classical snapping by doing the following:

16. Click the Editor menu and click Options.
17. Click the General tab and check Use classic snapping, and click OK to exit the dialog box

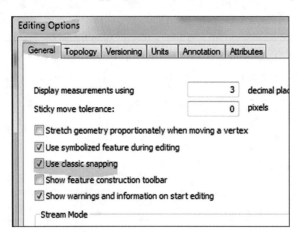

Comment: this method requires lot of time especially if there are many lines to be digitized and too much gaps.

Automatic Vectorization

This method will allow user to convert automatically, the image into vector. This approach is faster and reduces the time of processing.

18. Click the **Editor** menu on the *Editor* toolbar and click **Start Editing**.
19. Choose **Stream.shp** and click **OK**.
20. Click the **Vectorization** menu on ArcScan toolbar/point to **Vectorization Settings**

21. Click the **Vectorization** menu on ArcScan toolbar/point to **Vectorization Settings**
22. Change the **Intersection Solution** to **Geometrical**
23. Click **Style** change it to **Contour/OK**

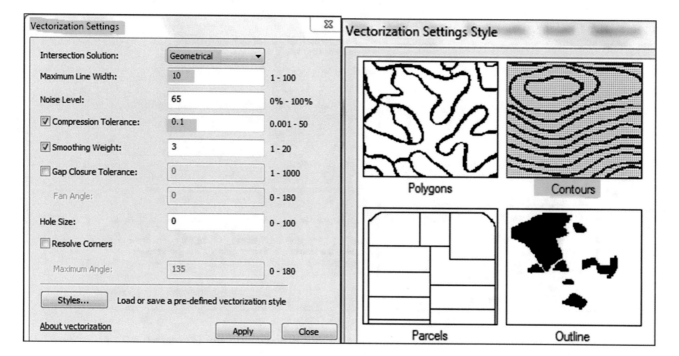

24. Change the **Maximum Line Width** value to 10.
25. Change the **Compression Tolerance** value to 0.1.
26. Click **Apply** to update the settings.
27. Click **Close** to the **Vectorization Setting** dialog box

Show Preview
28. Click the **Vectorization** in **ArcScan Tool** point to **Show Preview**
29. The **Show Preview** is displayed in the map

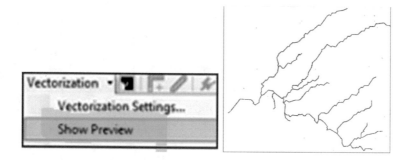

30. **Vectorization** in **ArcScan Tool** uncheck **Show Preview**

Note: if you see any think is not erased, use the previous steps and erase it.

Generate Features
31. Go to **Vectorization** menu and point to **Generate Features**
32. The **Stream** template became the active line feature template
33. Click OK

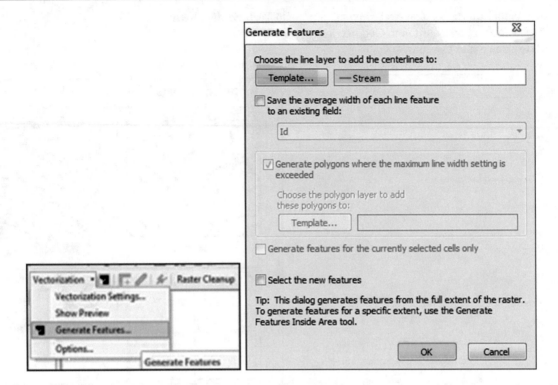

34. The Generate Feature command adds the centerlines to **Stream.shp** and the streams is generated

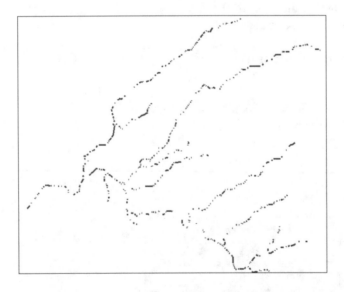

1. Go under **Editor** and select **Stop Editing**

Result: The stream is digitized.

Geodatabase

7

Introduction

ArcGIS works with different GIS and non-GIS file formats. Some were created by ESRI such as coverage, shapefile, geodatabase and non-ESRI products such as AutoCAD, MIF/MID, and others. The geodatabase is the built-in data structure for ArcGIS and is the primary data format used for editing and data management. Geodatabase combines "geo" (spatial data) with "database" to create a central data repository for spatial data storage and management. It can be leveraged in desktop, server, or mobile environments and allows you to store GIS data in a central location for easy access and management.

A Geodatabase in ArcGIS can be defined as a collection of diverse types of data kept in either Microsoft Access database or multiuser relational database (Oracle, Microsoft SQL Server, PostgreSQL, Informix, or IBM DB2) file systems. If it is held in a Microsoft Access database file system it is called a personal Geodatabase and if it is kept in a relational database file system it is called a file geodatabase. The differences between them are: (1) File geodatabases have unlimited storage space while personal geodatabases have a limit of 2 GB, (2) File geodatabases save their data in an ArcGIS folder format while the personal geodatabases save their data in the format of Microsoft access, and (3) File geodatabases can be compressed to save space.

The geodatabase in ArcGIS contains three primary dataset types: (1) Feature classes, (2) Raster datasets, and (3) Tables.

A Geodatabase can be generated from scratch by creating or collecting dataset types. After building a number of these fundamental dataset types, one can add to or extend their geodatabase with more advanced capabilities (such as adding topologies, networks, or subtypes) to model GIS behavior, maintain data integrity, and work with an important set of spatial relationships.

Geodatabases work across a range of database management systems (DBMS). Which include architectures and file systems that come in many sizes, and have varying numbers of users. They can scale from small, single-user databases built on files up to larger workgroup, department, and enterprise geodatabases accessed by many users.

Creating a Geodatabase

This section will allow you to capture data using digital data to create feature classes and store it in a geodatabase.
Scenario 1: You are working as a hydrogeologist for Water Authority and you are asked to create a file geodatabase and fill it with point, line, and polygon feature classes. You are also going to use the "**Image_Rectify.tif**" image as a source to capture groundwater wells as a point feature class, the fault as a line feature class, and plant as a polygon feature class.

Electronic Supplementary Material: The online version of this chapter (https://doi.org/10.1007/978-3-319-61158-7_7) contains supplementary material, which is available to authorized users.

W. Bajjali, *ArcGIS for Environmental and Water Issues*, Springer Textbooks in Earth Sciences, Geography and Environment, https://doi.org/10.1007/978-3-319-61158-7_7

Create File Geodatabase

1. Start ArcMap
2. Open Catalog window\connect to Ch07
3. R-click **Result** in **Ch07** folder/select New/File Geodatabase.

4. Change the name from New File Geodatabase.gdb to "**Dhuleil.gdb**"

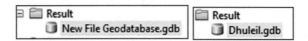

Create a Feature Class and Assign for it WGS84-GCS

To create a feature class, the file geodatabase must already exist. You are going to create **3-new feature** classes and call them **Well (point)**, **Fault (line),** and **Plant (polygon)**.

5. R-click on the "**Dhuleil.gdb**" and select New/Feature Class
   a. Name: Well
   b. Type: Point Feature
6. Click Next
7. D-click Geographic Coordinate Systems/scroll down and D-click World/scroll down and select WGS1984
8. Click Next/Next/Next/
9. Click Finish

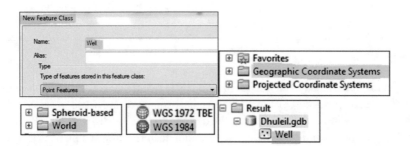

   Result: The Well feature class is created and it will be stored in the Dhuleil.gdb and added into TOC.

10. Repeat the previous steps and create a Fault (line) Feature Class and assign to it the GCS_WGS_1984

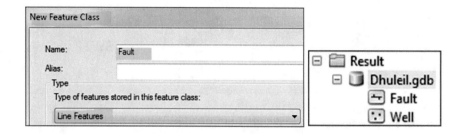

11. Repeat the previous steps again and create a Plant (polygon) Feature Class and assign to it the GCS_WGS_1984

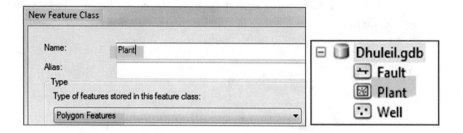

Result: The Fault and Plant feature classes are integrated into the TOC.

Capture the Feature Classes Using an Image

You are going to digitize the two Wells, the Fault, and the Plant using a rectified image. The image is called "**Image_Rectify.tif**" and it is registered in latitude-longitude and associated with Jordanian datum (D_Jordan).

12. Click Add Data/browse to \\Ch07\Data\Image folder and integrate "**Image_Rectify.tif**"
13. If you don't see the image/r-click "**Image_Rectify.tif**" it in the TOC and point to Zoom To Layer

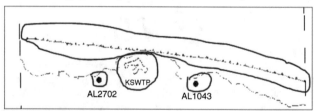

14. On the Editor tool/click on the drop-down arrow/Start Editing

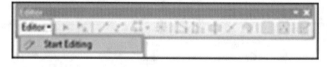

15. Click the Well feature template in the Create Features window

16. Click the Point tool **+· Point** under Construction Tools window
17. Click in the center of the well (AL2702) on the image "**Image_Rectify.tif**." then click on the second well (AL1043)

Result: The two points are created on the map and the second point is selected.

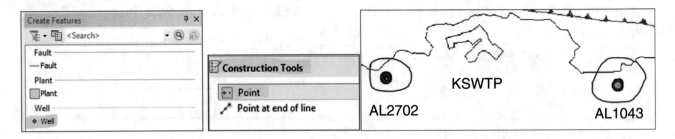

18. For the "Fault" layer make sure you select Fault feature template in the Create Features window

19. Click the Line tool  under Construction Tools
20. Rest your pointer over the endpoint of the existing fault in the eastern portion of the image and click once
21. Using the image as a guide, digitize the new line by clicking the map each place you want to add a vertex
22. Press F2 or double-click at the end of the fault to finish digitizing

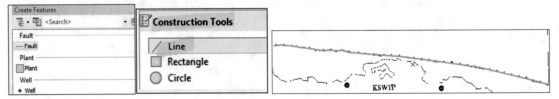

Result: The line is created on the map and it is selected.

23. Zoom In around the KSWTP in the "**Image_Rectify.tif**"

24. Click the Plant from the Create Features window and Polygon 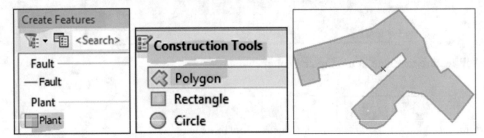 under the Construction Tools
25. Place your pointer over the right-east corner of the KSWTP and click once, then click along the outside edge of the KSWTP in the image to create the polygon
26. Press F2 or double-click at the end of the plant on the image to finish digitizing the plant

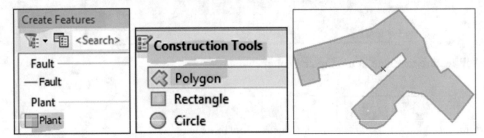

Result: The Plant is created on the map and it is selected.

27. On the Editor toolbar/click Editor drop-down/click Save Edits

28. Change the symbols of the Wells, Fault and Plant based on your taste

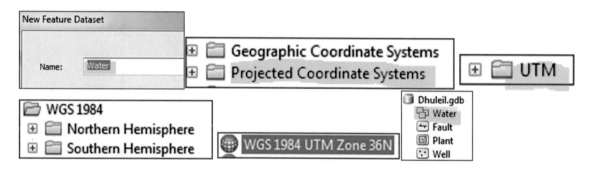

Create Feature Dataset and Import Shapefiles into It

29. Open Catalog window/browse to \\Ch07\Result folder
30. R-click "Dhuleil.gdb"/select New/Feature Dataset
31. Name: Water
32. Next/click Projected Coordinate System/select UTM/Select WGS1984/
33. Open Northern Hemisphere/select WGS1984 UTM Zone 36N
34. Click Next/Next/Finish

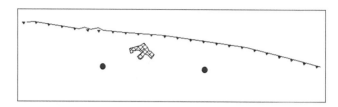

Result: The Water Feature Dataset is created in the "**Dhuleil.gdb**" and it is registered in UTM zone 36 N and associated with the WGS 1984 datum.

Important Note: Any file that will be imported into the Water Feature Dataset, will have its coordinate system converted into the coordinate system of the Water Feature dataset automatically.

Import the Stream, StudyArea, and Dam into the Water Feature Dataset

This step will import three shapefiles (Dam.shp, Stream.shp, and StudyArea.shp) into the Water Feature dataset. The original coordinate of the three shapefiles are in latitude and longitude and associated with the WGS 1984. Once these files are imported into the Water Feature Dataset, they will be converted into WGS_1984_UTM zone 36 N.

35. In the Catalog window/R-click Water Feature Dataset/Import/Feature Class (multiple)
36. Input Features/browse to \\Ch07\Data\Shapefile folder, highlight **Dam.shp**, **Stream.shp**, and **StudyArea.shp** and click Add
37. Click OK

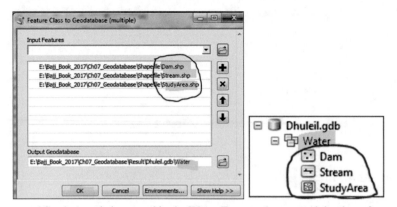

Result: The Dam, Stream, and StudyArea is imported in the Water Feature dataset and the three feature classes are displayed in the TOC.

38. Make the StudyArea Hollow, and change the symbols of the Dam and Stream feature classes as in the image below

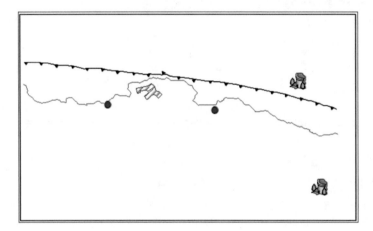

39. In Catalog window/R-click the Water Feature Dataset and find out what is the coordinate system?
40. In Catalog window/R-click the Dam feature class and find out what is the coordinate system? Repeat this for the Stream and StudyArea

Create a Relationship Class

A relationship class can be created between any feature class or table within a geodatabase using tools either in ArcCatalog or the Catalog window in ArcMap. These tools can be used to create simple, composite, and attributed relationship classes. In this section, you will create a relationship class between a well feature class and two tables. The first table contains water chemical analysis and the second table contains hydrological information about the wells. The relationship class will be created first between the well feature class and Table1 feature class based on the Well Inventory NO. (INVEN_No). The second relationship will be created between Table1 and Table2 also based on INVEN_No. Relationship classes support all cardinalities: one-to-one, one-to-many, and many-to-many. In this section one-to-one will be used. Relationship classes provide many advanced capabilities not found in ArcMap that joins and relates. A relationship class can be set up so that when the user modifies an object, the related objects update automatically. This can involve physically moving related features, deleting related objects, or updating an attribute.

Scenario 2: You are a hydrogeologist working in Jizi catchment area in Sultanate of Oman. You have been asked by your superior to perform an analysis dealing with the following files in ArcGIS:

1. Well.shp (groundwater wells)
2. Catchment.shp (Jizzi watershed area)

3. Table1.txt (table containing chemical water analysis from the wells)
4. Table2.txt (table containing hydrological information about the wells)

   In order to do this job you have to do the following: create a Personal Geodatabase, integrate the proper files into it and create a relationship class.

Create Personal Geodatabase

1. In ArcMap/Insert Data Frame and call it Jizzi
2. Open Catalog window/R-click \\Result folder/New/Personal Geodatabase
3. Rename the New Personal Geodatabase.mdb "**Jizzi.mdb**"

Integrate Shapefiles and Tables into Personal Geodatabase

4. R-click "**Jizzi.mdb**"/Import/Feature Class (Multiple)
5. Browse to \\Data\Jizzi folder/highlight Catchment.shp and Well.shp
6. Click Add then OK

Result: Catchment and Well feature classes are now imported into Jizzi.mdb.

7. R-click "**Jizzi.mdb**"/Import/Table (Multiple)
8. Browse to \\Data\Jizzi folder/highlight Table1.txt and Table2.txt
9. Click Add then OK

Result: Table1 and Table2 feature classes are now imported into Jizzi.mdb.

10. Add Well, Catchment, Table1, and Table2 feature classes into the TOC.

Create Relationship Class

First create Relationship Class in Jizzi.gdb between Well and Table1

11. In Catalog window/R-click **Jizzi.mdb**/New/Relationship Class
12. Name: Well_Table1
13. Origin table: Well
14. Destination table: Table 1
15. Next

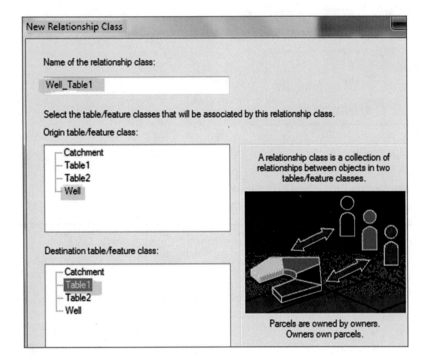

16. Select Simple (peer to peer) relationship

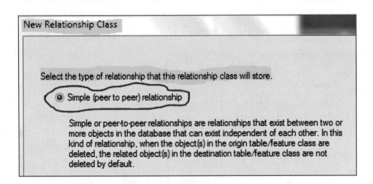

17. Click Next/Next/Select 1 – 1 (one to one)

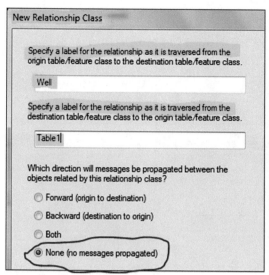
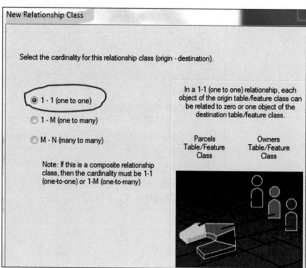

18. Click Next (select/No, I do not want to add attributes to this relationship class)

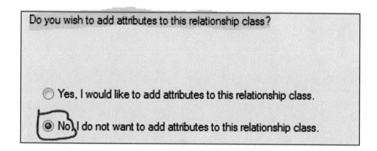

19. Click Next (select the Primary & Foreign key, which is INVEN_NO)

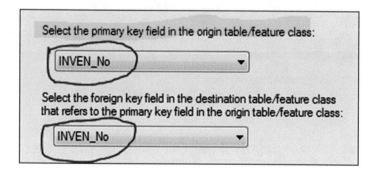

20. Click Next and Finish

> This is a summary of the relationship class:
>
> Name: Well_Table1
> Origin object class: Well
> Destination object class: Table1
> Type: Simple
> Forward Path Label: Well
> Backward Path Label: Table1
> Message propagation: None (no messages propagated)
> Cardinality: 1 - 1
> Has attributes: No
> Origin Primary Key: INVEN_No
> Origin Foreign Key: INVEN_No

Result: The Well_Table1 Relationship Class is now established.

Create Relationship Class in Jizzi.gdb Between Table1 and Table2

21. R-click **Jizzi.gdb**/New/Relationship Class
22. Name: 　　　　　　　Table1_Table2
23. Origin table: 　　　　Table1
24. Destination table: 　Table 2
25. Next

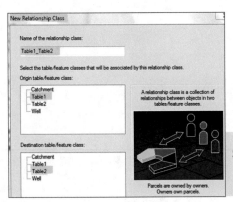

26. Next

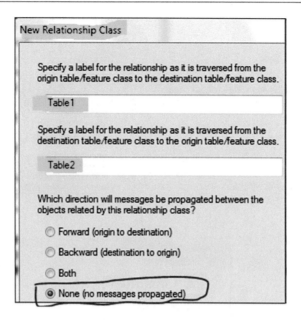

27. Next Select 1 – 1 (one to one)

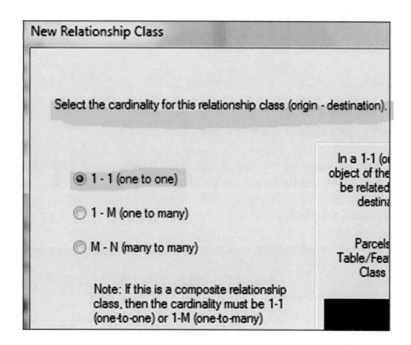

28. Check No, I do not want to add attributes to the relationship class

⊙ No, I do not want to add attributes to this relationship class.

29. Next

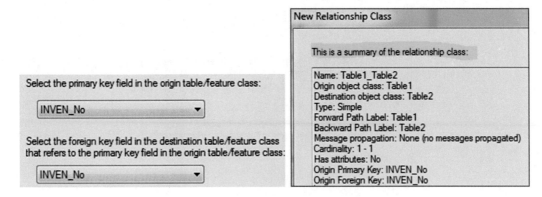

30. Click Finish

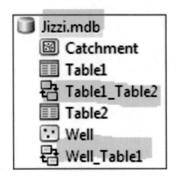

Result: The Table1_Table2 Relationship Class is now established.

Find the Wells that have Salinity Higher than 500 mg

31. In ArcMap make sure that the Well, Catchment, Table1, and Table2 feature classes are already in the TOC
32. If you don't see the tables, click the List By Source (above the table of content)

33. R-click Table1/Open/click Table Options [icon]/Select by Attributes
34. D-click TDS click > and type 500
    [TDS] > 500

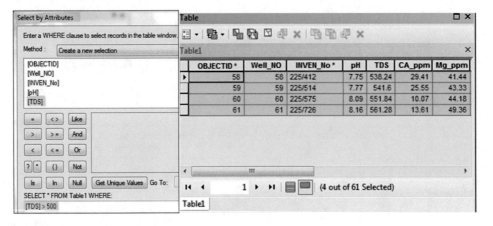

35. Click Apply/Close
36. Click Show Selected Records (at the bottom of the table)

Result: Four-wells are selected.

37. Click again Table Options/point to Related Tables/select Well_Table1: Table1

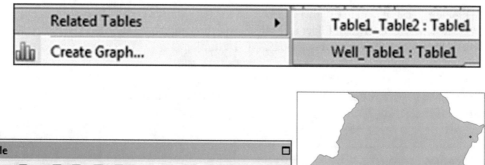

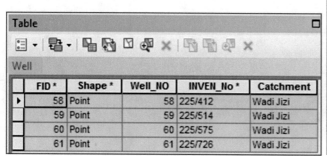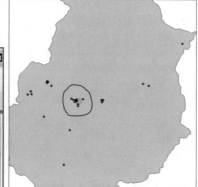

Result: Four-wells in the Well Attribute table are selected and these wells are selected in the catchment area.

38. Highlight the Table1

39. In Table 1/click again Table Options/Related Tables/select **Table1_Table2: Table1**

	OBJECTID *	INVEN_No *	Drilling	Elevation_m	Depth_m	Aquifer	Yield_m3_h
▶	53	225/412	Dec-87	115	500	Quaternary	43
	54	225/514	Jun-05	178	316	UeR	90
	55	225/575	Mar-02	175	311	UeR	200
	58	225/726	Jun-05	50	1126	Alluvium	90

Table2

Result: Four wells are selected and these wells are penetrating the following aquifers:

a. Quaternary: 1 well
b. UeR: 2 wells
c. Alluvium: 1 well

Projection and Datum Conflict

In the datum conflict in Chap. 5, you encountered a problem when you projected the well feature class from Jordanian Transverse Mercator into the Palestine Transverse Mercator. In order to perform the projection, the Jordanian datum had to be converted into WGS 84 and then the WGS 84 converted into the Palestine datum. The result was not a perfect projection as the output projected file didn't align correctly with the original file that registered in the Palestine Transverse Mercator. The discrepancies between the same well from the two files was around 177 m. To solve these issues the projection was performed in the geodatabase environment.

GIS Approach

40. Insert Data Frame and call it Datum Conflict
41. Create File Geodatabase in the Result folder and call it Projection.gdb
42. R-click Projection.gdb/New/Feature Dataset and call it "**Datum**"

43. Next/click the Add Coordinate System drop-down arrow ▣ ▾ icon
44. Click Import browse to \\Projection folder and select Well_PTM.shp and click Add
45. Click Next/Next/Finish

Result: The Datum Feature Dataset is created.

46. R-click Datum Feature Dataset/Import/Feature Class (Multiple)
47. Browse to \\Data\Projection folder and select Dhuleil.shp, Well_JTM, and Well_PTM.shp
48. Click Add/OK

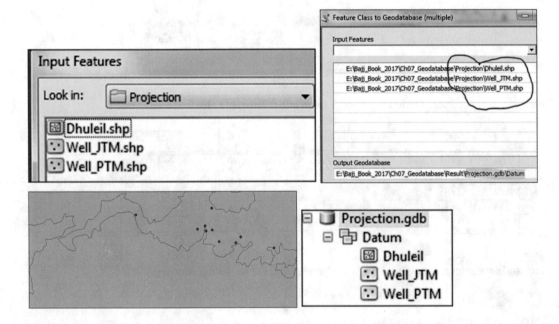

Result: The three feature classes **Dhuleil**, **Well_JTM**, and **Well_PTM** are reside in the Datum Feature Dataset and integrated into the TOC. The **Well_JTM**, and **Well_PTM** are both projected into the Palestine Transverse Mercator and they are aligned together perfectly.

Introduction

During the data creation, either using a tablet or on-screen digitizing can generate errors. The error can be due to human error, such as missing a point, line, polygon, or digitizing extra features. Errors can also be generated during scanning, tracing or during the georeferencing. An ArcGIS user can edit various types of data such as: feature data stored in shapefiles, geodatabases, and different tabular formats. The editing can include points, lines, polygons, and text.

Editing occurs in an edit session where you can create or modify vector features or tabular attribute information. Start an edit session when you are ready to begin editing, but remember to end the edit session when you're done. If you have more than one data frame in your map, you can only edit the layers in one data frame at a time, even if all data is in the same workspace. The editing of the data can be done if they are either in the same or in different coordinate systems.

Topology is an advance way to edit the data and it is define as a data structure that creates connections and describes the spatial relationship between point, line, and polygon features. In other ways, the topology is simply the arrangement of how the 3-different features (point, line, and polygon) share geometry. All spatial elements in a GIS layer are connected in some fashion to each other which allows the layer to be categorized, queried, manipulated, and stored more efficiently. The topology is also a set of rules, behaviors, and models on how points, lines, and polygons share coincident geometry. For example, two adjacent catchment areas will have a common water divide between them which they share. The set of sub-catchment polygons within each watershed must completely cover the watershed polygon and share edges with the whole catchment boundary. The topology is a useful data structure concept in GIS which allows GIS users to know: the location of the feature, what is connected to it, what is surrounded by it, and how to identify spatial relationships with other features. It can also help to get around using the nodes and vertices to accomplish various spatial analysis tasks. In GIS, one can find, and trace, a route on a map between two cities, and measure the distance and time of arrival.

When topology is applied in GIS, a data structure table is built from _nodes_ and _chains_ of the features. The tables are used to determine various relationships such as: what is connected, what is adjacent (left and right), and what is the direction of the chains. Topology is applied after digitizing and editing. When data is digitized or created there is no connection, or relationship, to the feature that has recently been digitized. This means that no informational content associated with point, line, or polygon is available, except location. For example, if you digitize a river, and its tributaries, then run the topology, it will build the spatial information. It does this by recognizing the nodes at the end of each digitized stream and creating new nodes at intersections where the river crosses. The end result is that each segment of the river consists of three topological chains separated by a node (figure below). One stream segment consist of Arc 1, Arc2, and Arc 3 using start node and end node. Arc 1 has node 1 as its starting node and node 4 as its ending node.

Topology offers special information to the data structure, provides powerful functions for spatial analysis and presents a number of advantages to GIS. The topology allows users to calculate the spatial information and property of the features. The spatial property for the point is location (X, Y), the line is the length, and the polygon is the perimeter and area. Topology provides spatial relationships which allow users to query the data and provide spatial analysis when running the network analyst.

Electronic Supplementary Material: The online version of this chapter (https://doi.org/10.1007/978-3-319-61158-7_8) contains supplementary material, which is available to authorized users.

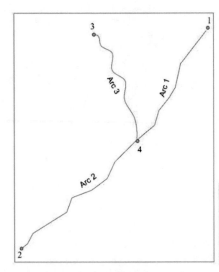

Arc	Start Node	End Node
Arc1	1	4
Arc2	4	2
Arc3	3	4

In this chapter, you are going to be introduced to various types of editing, which range from simple, advance, and all the way to topological editing.

I. Simple Editing
   a. Delete
   b. Move
   c. Split
   d. Reshape
   e. Modify
   f. Merge
II. Advance Editing
   a. Overshoots and undershoots
   b. Generalize feature
   c. Smooth feature
III. Topological Editing Using Geodatabase
   a. Fix Lines using topology
   b. Fix polygons using topology

I: Simple Editing

Simple Editing means a variety of basic editing that can be performed on a point, line, and polygon features. Editing allows you to use some commands to perform certain functions such as Delete and Move, or using the Editing Tool to perform several type of duties. The Editor toolbar includes several commands that help users to edit their data. It also allows you to start and stop an edit session, access a variety of tools, have commands that create new features or modify existing ones, and

can save your edits. To edit data, you need to add the Editor toolbar to ArcMap by clicking the Editor Toolbar button on the Standard toolbar or access it through the Customize menu/Toolbars/Editor.

The Editor toolbar executes quite a few commands such as trace, cut, reshape, split, rotate, and many more functions.
Scenario 1: You are giving a shapefile that was digitized from an aerial photograph and your boss asked you to modify it by deleting and moving some polygons.

Delete Function

In this step you are going to delete 2-polygons: H and G from the Farm layer

1. Start ArcMap and rename the Layers Data Frame "**Editing**"
2. Integrate the **Farm.shp** from \\Data\Q1 folder
3. Right-click **Farm.shp** and point to Label Features
4. Click Editor toolbar/Start Editing/highlight Farm/Continue

5. Click Edit Tool ▸ in the Editor Tool
6. Point to polygon H and click it, the H polygon will be selected
7. Click the Delete key on the keyboard
8. The H Polygon will be deleted
9. Repeat the previous steps to delete polygon G

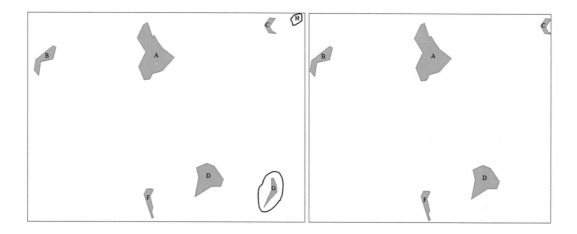

Move Function

In this step you are going to move 2-polygons: D and F.
   You have found that the actual location of the land D is between lands A and B. The location of land F is between lands A and C. So you have decided to move them into their correct locations.

10. Make sure you are in editing mode
11. Click Edit Tool in the Editor Tool
12. Point to polygon D and click it, the D polygon will selected
13. Click on it again and drag it between lands B and A
14. Repeat point 12 and 13 and drag polygon F between A and C then click on an empty place to deselect the F feature

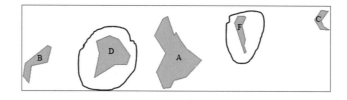

Split Function

The piece of land A is big and used for cultivating the potato crop. You have decided to split it into two parts and use it to cultivate two products: tomato and potato.

15. Make sure you are still in editing mode
16. Zoom to polygon A
17. You will split the polygon A between the points 1 and 2 as seen in the figure
18. Editor/Snapping/Snapping Toolbar

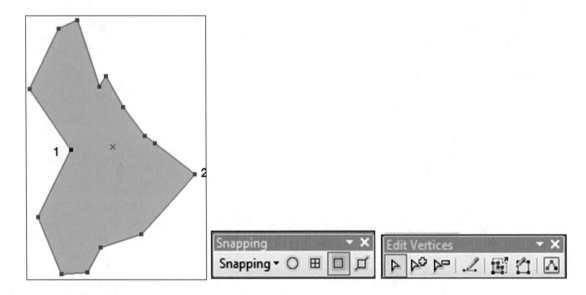

19. In the Snapping Toolbar/click the Snapping drop down menu
20. Make sure the "**Use Snapping**" icon is checked
21. Click Vertex Snapping (third icon on Snapping Toolbar)

22. Click Edit Tool [▶] in the Editor Tool
23. Point to polygon A and double click it, the A polygon will be selected

   Result: Polygon A will have one node (red) and 14-vertexes (greens).

24. Highlight the Farm in the Create Features Dialog Box

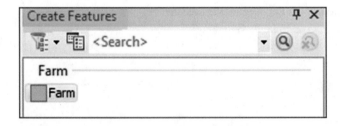

25. Click "**Cut Polygons Tool**" [⊞] on the Editor Tool
26. Point you cursor toward point 2, it will snap
27. Then point your cursor toward point 1, it will snap, then double click it

28. The polygon A will split into two polygons and both have the label A
29. Open the attribute table of **Farm** layer
30. You will see two records selected and both of them labeled "A"
31. Highlight the Lower Label A, replace it by typing **M**
32. Then hit Enter/close the Farm Attribute Table

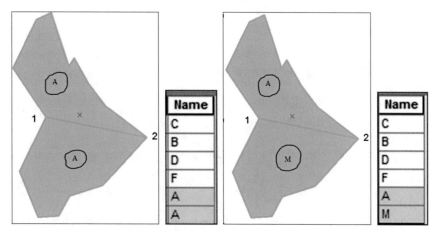

Result: The polygon A is now split into two polygons A and M.

Reshape Function

The piece of the **Farm** that has a label B, is now going to be improved and expanded to make more land available for agriculture. The Farm with the label B is going to be reshaped and modified to fit the size and shape of the **LandB;** which has a rectangular shape.

33. Add **LandB.shp** from \\Data\Q1 folder to the Editing Data Frame
34. Right click the LandB layer in the TOC/Zoom To Layer
35. In the TOC, click the symbol of the **LandB** layer, then click Hollow and make the Outline Width = 1 Click OK to exit the Symbol Selector dialog box

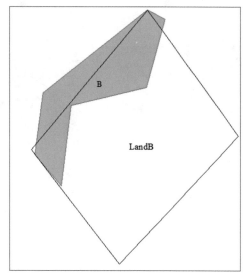

36. Right click **LandB** layer in the TOC/Edit Features/click the Organize Feature Templates
37. The Organize Feature Templates dialog box should be displayed
38. Click LandB and point to New Template

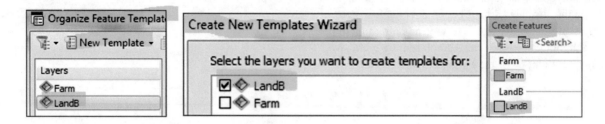

39. The Create New Template Wizard should be displayed
40. Highlight LandB and click Finish
41. Close the Create New Template Wizard dialog box display
42. The LandB is added into the Create Features Dialog

   Result: The two layers **LandB** and Farm are now in the editing mode. However, you are going to reshape only the land B of the **Farm** layer.

43. In the Snapping Toolbar/click Snapping
44. Make sure "**Use Snapping**" is checked and the Vertex Snapping (third icon) is highlighted
45. Click Vertex Snapping (third icon on Snapping Toolbar)
46. Click Edit Tool in the Editor Tool
47. Double click land B of the **Farm layer**
48. The polygon will have one node (red) and all of the vertexes (greens)
49. Click Edit Tool in the Editor Tool and place it on vertex 1
50. Drag it and place it on the left upper corner of the **LandB** close to vertex 7 (as seen in the sketch)
51. Click vertex 6 and drag it to the lower left corner of the **LandB**
52. Click vertex 3 and drag it to lower right corner
53. Place your Edit Tool above vertex 4, right click and delete the vertex
54. Repeat the previous steps and delete vertex 5 and 7
55. Click Edit Tool somewhere outside the drawing
56. The farm with label B will be reshaped and will fit the rectangular shape of LandB
57. Go to Editor and click Stop Editing /click Yes
    In the TOC, r-click the Farm layer/Zoom to Layer

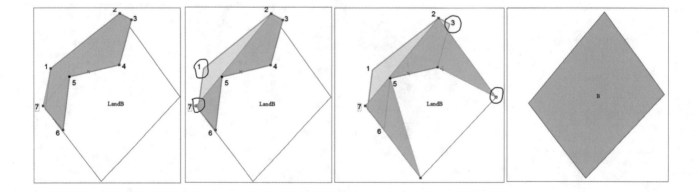

Modify Feature

One of the ways to modify a feature in ArcMap is to add features through digitizing and then update the attribute table.

Scenario 2: The geology of Dhuleil was subject to a detail study to update its outcropping formations. A group of geologists went to the field and, with the use of GPS, they delineated the outcropping formation that was missing from the old map. As a GIS technician, your duty is to use the new data to update the original geological map "**Geology.shp**".

1. Insert a new Data Frame and call it **Geology**
2. Integrate the **Geology.shp** and **Field_Geology.shp** from \\Data\Q2 folder
3. Right click **Geology.shp** and point to Properties/Symbology/Categories/Unique values
4. Go to the Value Field "GEOLOGY"/Click Add All Values/Uncheck all other values
5. Click OK
6. Right click the **Field_Geology.shp** point to Label Features

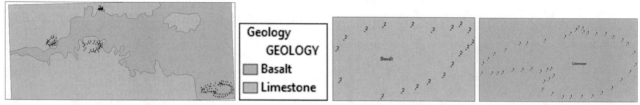

Result: The Geology layer will be displayed in 2 classes (**Basalt & Limestone**) and the Field_Geology layer is now labeled.
Note: The next step is to digitize the basalt and limestone features using the Field_Geology layer as a reference. Labels 1 and 2 represent the limestone and the basalt formations respectively. The captured limestone is outcropping above the basalt, and the outcropping basalt is above the limestone.

7. Editor/Start Editing/Editor/Snapping/Snapping Toolbar
8. In the Snapping Toolbar/click Snapping and make sure the "**Use Snapping**" is checked
9. Click Point Snapping on the Snapping toolbar (first icon on Snapping Toolbar) to make it active

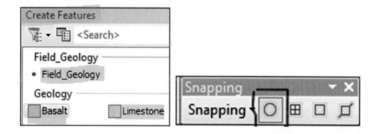

10. Zoom in around the Field_Geology layer in the lower right corner of the **Geology** layer
11. In the Create Features click the Limestone class under geology
12. In the Construction Tools click the Polygon Symbol

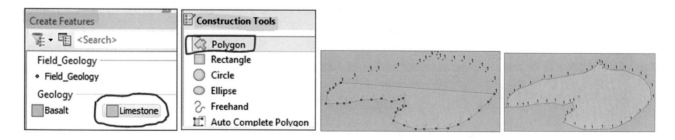

13. Click one point of the **Field_Geology.shp**
14. Then click a second point, and continue till you finish all the points
15. When you reach the last point double click to finish digitizing
16. Repeat the previous steps to finish digitizing the rest of the limestone formations

17. In the Create Features, click the Basalt class under geology
18. In the Construction Tools click the Polygon Symbol
19. Zoom in around the basalt points in the center of the **Geology.shp**
20. Click one point of the **Field_Geology.shp**
21. Then click a second point, and continue till you finish all the points
22. When you reach the last point double click to finish digitizing
23. Repeat the previous steps to finish digitizing the basalt formations
24. Open the attribute table of the **Geology.shp**
25. You will notice that all the limestone and basalt formations are added

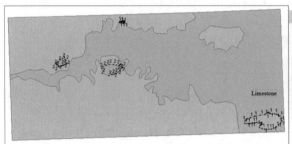

Geology

FID	Shape *	DHUL2_	DHUL2_ID	GEOLOGY	Area	Perimeter
0	Polygon	1	1	Basalt	230023000	147809
1	Polygon	2	2	Limestone	11807200	16254.3
2	Polygon	3	3	Limestone	82319400	66309.5
3	Polygon	4	4	Limestone	180966000	98912.6
4	Polygon	0	0	Limestone	0	0
5	Polygon	0	0	Limestone	0	0
6	Polygon	0	0	Basalt	0	0
7	Polygon	0	0	Basalt	0	0

Update the Area and Perimeter Field in the Geology Attribute Table

26. Go to Editor, Stop Editing/click yes to save your edits
27. Right click the Area field in the attribute table/Calculate Geometry/click Yes
28. The Calculate Geometry dialog box should be displayed
29. Make the Property: Area
30. For the Coordinate System: check the "Use Coordinate system of the data source" WGS1984 UTM Zone 36N
31. Make the Unit: Square Meters
32. Click OK/Yes

Result: The Area of the new digitized area will be calculated.

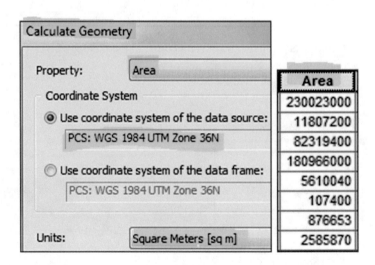

33. Right click the Perimeter field in the attribute table/Calculate Geometry/click Yes
34. The Calculate Geometry dialog box should be displayed
35. Make the Property: Perimeter

36. Coordinate System: (Use Coordinate system of the data source) WGS1984 UTM Zone 36N
37. Make the Unit: Meters
38. Click OK/Yes

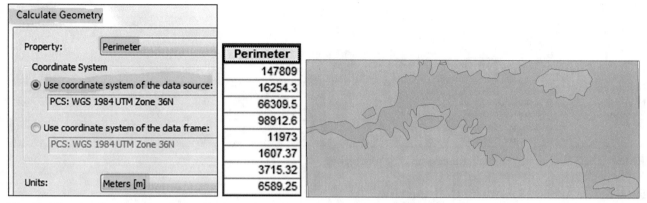

Result: The Attribute Table of the **Geology** layer is updated and now contains eight records: three basalts and five limestones.

Merge Function

The merge function works with a single layer, and it can group selected records of a line or polygon features into one feature. In this scenario you are going to reduce the numbers of basalt and limestone records into one record each. So instead of having five records for the limestone in the attribute table, you will have only one record; same thing for the basalt features.

39. Insert Data Frame and call it **Merge**
40. Integrate **Geol_Dhul.shp** from the \\Ch08\Data\Q2 folder
41. Classify the **Geol_Dhul** layer based on the GEOLOGY field as shown in the previous section
42. Open the attribute table of the **Geol_Dhul** layer
43. The table consists of 8 records; **5 limestone records** and **3 basalt records**

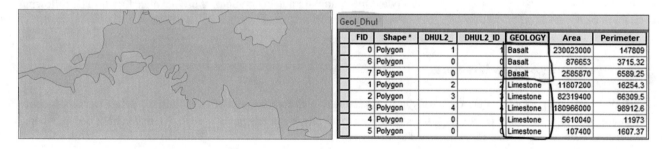

44. Right click **Geol_Dhul** layer/Edit Features/Start Editing
45. Open the Attribute Table of **Geol_Dhul** layer
46. Click Table Options and point to Select By Attributes
47. Write the following SQL statement: **"GEOLOGY" = 'Limestone'**
48. Click Apply then Close

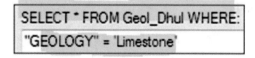

Result: The 5-records of the limestone are selected.

49. Click Editor toolbar then select Merge
50. Select the first record "**Limestone (Geol_Dhul)**"
51. Click OK

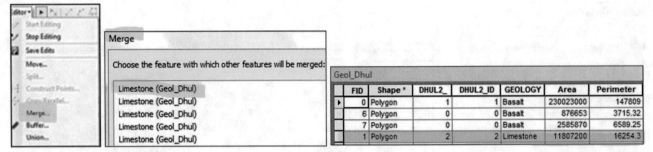

Result: The 5-limestone records have now become one record.

52. Open the Attribute Table of the **Geol_Dhul** layer
53. Click Table Options then click on Select By Attributes
54. Write the following SQL statement: "**GEOLOGY**" = '**Basalt**'
55. Apply/Close
56. Go to the Editor/Merge
57. Select the first record "**Basalt (Geol_Dhul)**"
58. Click OK

  Result: The attribute table of the **Geol_Dhul** layer consists now of only two records.

59. Go to the Editor/Stop Editing and exit ArcMap

II: Advance Editing

In this section guide you will learn how to use the **Advanced Editing** tool in ArcMap to edit existing GIS features, how to fix some common digitizing errors, and how to update the spatial data using some of these advanced tools. The following topics will be covered:

Fixing Overshoots and Undershoots

Overshoots and undershoots are very common digitizing errors that affect the quality of the digitized data. Overshoots occur when a line that is supposed to terminate at the edge of another feature extends past the edge. An undershoot occurs when a line doesn't reach the edge where it is supposed to terminate.
    The overshoot is fixed by trimming it to a selected edge, while the undershoot is fixed by extending it to a selected edge.

GIS Approach
1. Start ArcMap/File/Open and browse to \\Data\Q3 folder and Double click **Editing.mxd**

    Result: A map display showing the aerial photograph of St. Louis County, Streets, Rivers and Lakes layers.

2. Customize menu/Toolbars/Advanced Editing

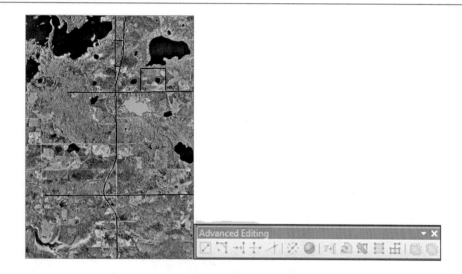

Result: Advanced Editing toolbar is displayed, but none of the tools are active. The tools will become active once you've:

a) Started an edit session
b) Selected a feature to edit it

3. Bookmarks menu/point to Undershoot

   Result: The streets display and the ends of the square street are not connected to the main street.

4. In the TOC, right click **Street_MN**/Edit Features/Start Editing

   Result: Some of the Advance Editing tool are now active.

5. Click Edit tool in the Editor Toolbar
6. Click the street that you want the undershoot street to connect to.

7. From the Advanced Editing toolbar, click the Extend tool (third icon from left)
8. Then click on one of the undershoot streets.

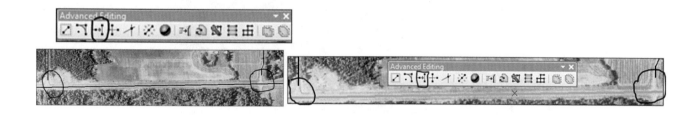

9. The undershoot street extended to the selected street centerline.
10. Next extend the other undershoots as well (as in the previous step).
11. Zoom out to the full extent and clear the selected features

Correct Overshoots of the Street

12. Bookmarks menu/point to Overshoot
13. Notice the overshoots where two small streets, inside the loop, extend beyond the straight street
14. Make sure you are still in editing mode
15. With Edit tool, select the straight street that you want the two overshoot streets to be connected to

16. Now click the Trim tool 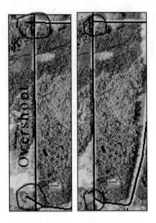 (4th icon from left).
17. Click the end of the overshoot on one street.
18. Both overshoots are trimmed off
19. Zoom out to the full extent and clear the selected features

Generalize a Stream Feature

Generalizing reduces the number of vertices that describe a feature, so the feature's shape is somewhat less precise. The step will allow you to reduce the number of vertices that describe the **River_Mn.shp**.

1. Bookmarks menu/River
2. Make sure the River_MN.shp is selected in the "Create Features"

3. Click the Edit tool and then double click the River_MN layer to display its vertices.

4. On the Advanced Editing toolbar, click the Generalize tool (2nd icon from right).
5. For the Maximum allowable offset, enter 5, then click OK

6. Double-click the River_MN layer again with the Edit tool to see its vertices.

Result: The number of vertices has been significantly reduced. If you zoom in closer at any part of the stream, you can see a slight effect on the shape of the stream.

7. Zoom out to the full extent and clear the selected features

Smooth a Lake Feature

The outline of the lake next to the stream feature appears rough because it was digitized with too few vertices. To improve the appearance of the lake, you will use the Smooth tool.

8. Bookmark/Sunshine Lake
9. Make sure the **Lake_MN** layer is selected in the "Create Features".
10. Double-click the lake with the Edit tool to display its vertices.

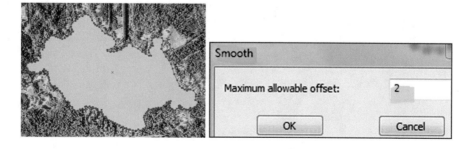

11. On the Advanced Editing toolbar, click the Smooth tool 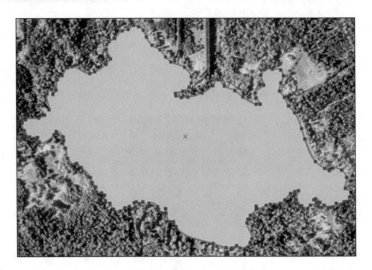 (last icon).
12. For the Maximum allowable offset, enter 2, then click OK.
13. Now the lake outline is much smoother.
14. Double-click the lake feature again to display the vertices.
15. The border of the lake looks smoother now.

16. Zoom out to the full extent and clear the selected features
17. Stop Editing/click Yes (or NO, in case you want to practice)/Exit ArcMap

III: Topological Editing Using Geodatabase

Geodatabase topology is a set of rules that define how the features in one or more feature classes share geometry. The topology can be created in the Catalog window or ArcCatalog and can then be added to ArcMap as a layer to be edited. The topology rules allows a user to identify the topology errors that are present. For example a line (rivers, faults, and roads) might have a dangle, where one end of the line is not connected to another line, which are errors that you need to fix. The ArcGIS also allows you to validate the geodatabase topology to see if the edits has been following the topology's rules correctly.

Fix Fault System Using Topology

The catchment area of Wadi Andam-Halfyan in the Izki region in Oman has many wells. They are used mainly for domestic water supply and agriculture. The area has revealed a structural style that may have had a profound effect on the geomorphological and hydrological setting of the area. Two major fault trends have been observed; Fault A and Fault B are oriented north-east and north respectively, while Fault C oriented north-west. A detail field geological study of "**Fault A**" reveals that the fault actually consists of one section; not from different five segments. Therefore, Fault A should be corrected by joining the five segments into one fault system. You are going to use the Topology tool to carry out this job.

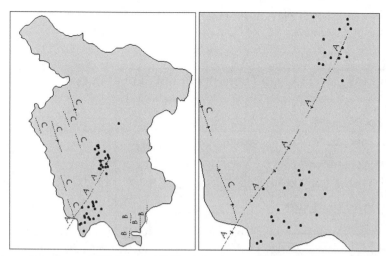

Note: The fault shapefile is projected onto UTM Zone 40N and the datum is WGS84 (**WGS_1984_UTM_Zone_40N**).

The figure above shows that the Fault feature A contains gaps, and your duty is to fix them using the Topology technique in Geodatabase environment. To fix the errors there are various editing approaches using the topology, and your duty is to implement two topology techniques.

Editing Using Topology

In order to use the topology to fix the errors in the database you must do the following:

1. Create a Geodatabase and a Feature Dataset
2. Integrate the Fault into the Feature Dataset as a feature class
3. Build a Topology and set the Topology Rules

Create Geodatabase

1. Launch ArcMap and rename the Layers data frame "**Fault**"
2. Open Catalog window and browse onto \\Ch08\Result folder, R-click Result/New/choose File Geodatabase
3. Enter "**Resources**" as the name of the New File Geodatabase.gdb
4. Right click **Resources.gdb**/New/Feature Dataset
5. Make the Name: Water
6. Click Next and open the "Projected Coordinate Systems"/UTM/WGS1984/
7. Select Northern Hemisphere then select "WGS 1984 UTM Zone 40 N"
8. Click Next/Next/Finish

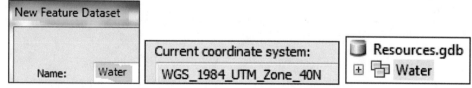

Result: The Water feature dataset is created.

9. Right click on the Water feature dataset/Import/Feature Class (single)
10. Input Features: \\Data\Q4\Fault.shp
11. Output Location: \\Data\Resources.gdb\Water
12. Output Feature Class: Fault

13. Click OK (The Fault feature class displays in the TOC)
14. Right click Fault in the TOC/Remove

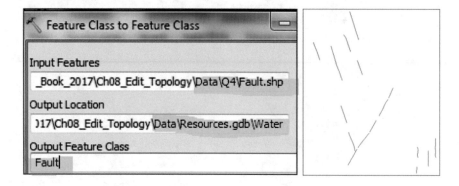

Building Topology and Set the Rules

15. Right click the Water Feature Dataset/New/Topology/Click Next
16. Call it Fault_Topology and accept the cluster tolerance (0.01 m)
17. Click Next/Select Fault/Next and accept the default/Next
18. Click Add Rule
19. Make the features of the feature class: Fault
20. Rule: Must Not Have Dangle
21. Click OK/Next/Finish
22. Click Yes to validate the topology

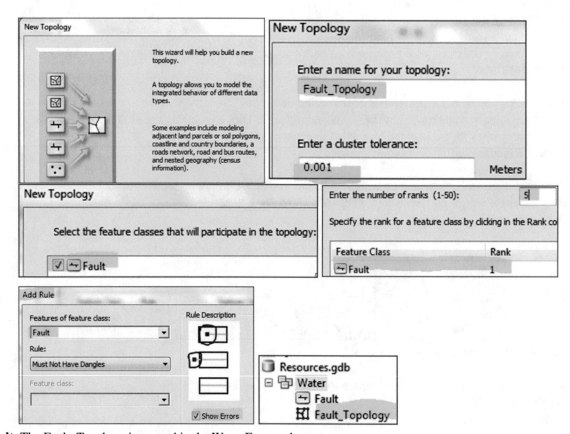

Result: The Fault_Topology is created in the Water Feature dataset.

23. Right click Fault_Topology/Properties/Click the Errors Tab
24. Click Generate Summary
25. You will see 32 Dangles
26. Click OK

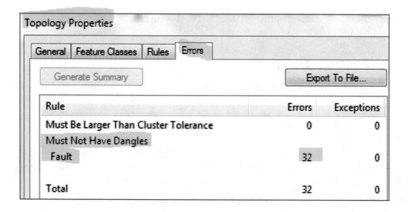

Fixing Dangles in ArcMap Using First Topology Approach

27. Drag the Water Feature Dataset into ArcMap from the Catalog window
28. The Fault_Topology and the Fault feature class will be added to the TOC.
29. In TOC, r-click Fault layer/Label Features

Comment: The 32 dangling nodes at the perimeter of the three fault systems (A, B, and C) is displayed. The goal is to fix the dangle in Fault A.

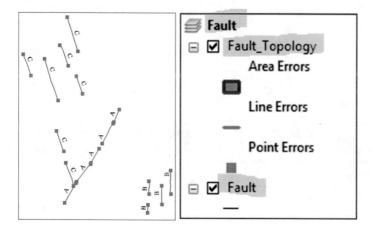

Topology Tool

30. Click the customize menu/Toolbars/click Topology tool
31. The Topology Toolbar will be displayed (none of the tools are active)
32. Click the Editor toolbar and click Start Editing
33. Some of the tools in the Topology toolbar will be active

34. On the Topology toolbar, choose "Select Topology"  (first icon)
35. Select "Geodatabase Topology"
36. Click OK
37. Zoom in between the two fault segments at the upper part of Fault A
38. Measure the distance between the two vertexes in meters in the upper two segments (round up)

Note: Here is the distance: 150.3 m.

39. On the Topology toolbar, select the "Fix Topology Error Tool" (second icon from last)
40. Click on a node, it will turn black/then right click on it and choose "snap"
41. Use the measured value from earlier as the snap tolerance value (here, 151 m)/Enter
42. The node from the upper fault segment will snap to the node of the lower segment

43. Click Validate Topology on Current Extent
44. The node will disappear
45. Zoom to Full Extent

   Result: The two sections of the A fault become one line segment.

46. Zoom into the next upper dangle error and measure the distance

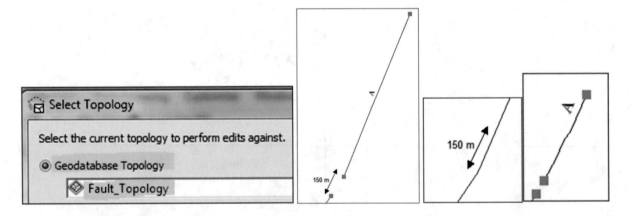

Result: The distance is 509.12 m.

47. Repeat the above steps in the same order to snap the two-nodes together

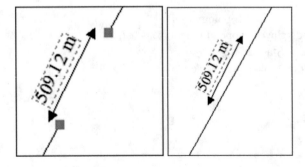

Fixing Dangles in ArcMap Using Second Topology Approach

This is a second approach to remove the gaps between the nodes along the fault segments based on the topology approach. To proceed, you need to make sure that you are in editing mode.

48. Zoom into the lower Fault A and measure the distance between the two nodes.

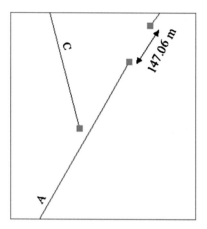

Comment: The distance is 147.06 m

49. Click Editor in the Editor toolbar
50. Click Snapping/Snapping Toolbar
51. Make sure the Point Snapping is highlighted
52. Click Create Feature on the right tab
53. Highlight Fault in the Create Feature dialog box
54. Click Line under Construction Tools

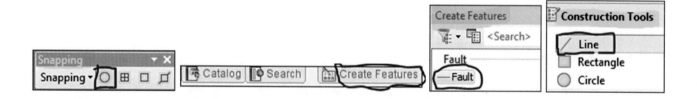

55. Click the Straight Segment ▱ Tool in the Editor toolbar
56. Right click the upper node/Snap To Feature/Endpoint
57. Move your cursor to the node you want to connect the line
58. Right click on the node/Snap To Feature/Endpoint/
59. Right click on the node once again/click Finish Sketch

   Result: A cyan line is created and connects the two nodes.

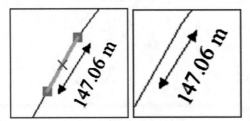

In order to connect the cyan line between the two nodes to the original fault, you should validate the topology.

60. On the Topology toolbar click the Validate Topology In Current Extent

Result: The cyan line and the two-nodes will disappear.
Continue correcting the gaps along Fault A

61. Click Stop Editing under the Editor dropdown menu when you finish

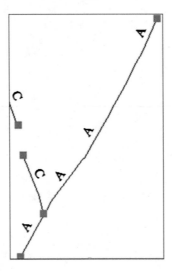

Fix Watershed Using Topology

Two catchment areas were created using the Hydrology tool in the Spatial Analyst. Two different thresholds were used and the result two catchment areas were created. The two catchments don't cover each other and in this exercise, you will use the Topology Tool to make them identical to each other.

62. Insert New Data Frame and call it Watershed
63. In the Catalog window right click \\Result\Resources.gdb\New\Feature Dataset
64. Make the Name: **Shed**
65. Click Next/then in the New Feature Dataset dialog box

66. Click the drop down arrow of the Add Coordinate System ⊕ ▾ and click Import
67. Import the coordinate from **Watershed_1.shp** from \\Ch08\Data\Q5 folder

The coordinate System of the **Watershed_1.shp** is "NAD_1983_UTM_Zone_15N"

68. Click Next/Next/Accept the Default/and click Finish
69. Right click on Shed/Import/Feature Class (multiple)

70. Add **Watershed_1.shp** & **Watershed_2.shp** from \\Data\Q5 folder/
71. Click OK

Result: **Watershed_1.shp** and **Watershed_2.shp** will be displayed in ArcMap.

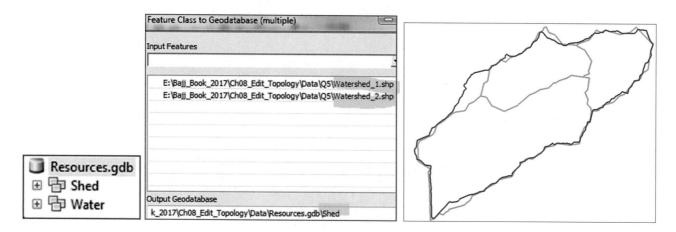

72. Remove the **Watershed_1** and **Watershed_2 layers** from TOC of Watershed in ArcMap

Build Topology Rule to Make Two Watershed Layers Cover Each Other

73. In Catalog window, right click on Shed feature Dataset/New/Topology
74. Click Next
75. Enter a name for your topology: Shed_Topology
76. Accept the default of the cluster tolerance/Next
77. Check Watershed_1 & Watershed_2
78. Click Next

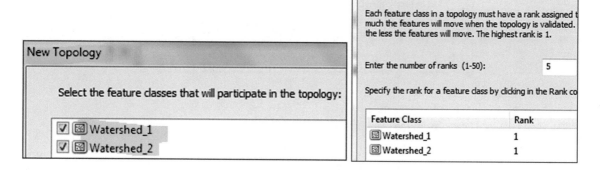

79. Accept the default/Next
80. Click Add Rules
81. Select Watershed_2.shp
82. Select "Must Cover Each Other"
83. Feature Class: Watershed_1.shp

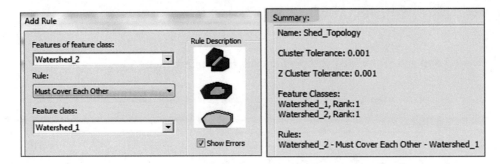

84. OK/click Next/Finish
85. Click Yes to validate the topology

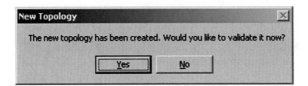

Result: The Shed_Topology is established and, you can see the detail of the topology rule in the Catalog window. The summary shows that the area is not covered by the red area.

86. In Catalog window/Right click Shed_Topology/Properties/click Error tab
87. Click Generate Summary
88. Click OK

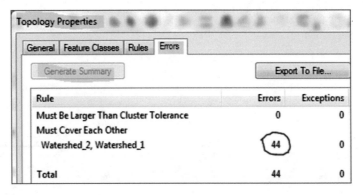

Result: There are 44 errors.

89. Drag the Shed Feature Dataset from the Catalog window into ArcMap

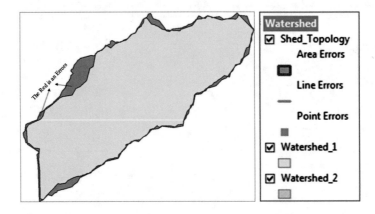

Result: The Shed_Topology, the Watershed_1 and Watershed_2 feature classes are added to the TOC.

90. Click Editor toolbar/Start Editing

Note: Make sure that the Topology toolbar is available in ArcMap.

91. Click the Select Topology tool 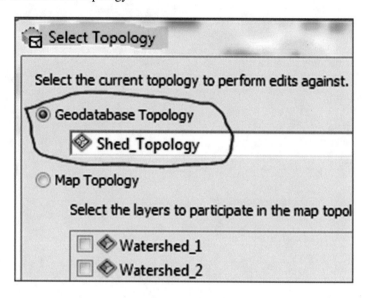 on the Topology toolbar
92. Click the circle next to Geodatabase Topology

93. Click OK

94. Click the Fix Topology Error tool 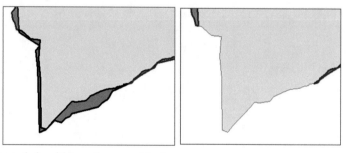 on the Topology Toolbar
95. Use the Fix Topology Error tool by dragging a box around the error in the lower-left corner

Result: The area that is selected should turn the error black.

96. Right click in the error and click Subtract
97. Repeat the above steps to fix the rest of the errors OR
98. Click the Fix Topology Error tool/drag box around Watershed_1 and Watershed_2
99. Right click in the error and click Subtract

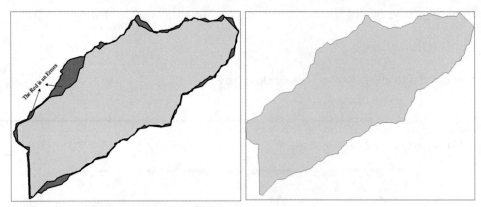

Result: The two watersheds become identical.

100. Click Editor and Stop Editing
101. Click Yes to save edits

Geoprocessing

9

Introduction

In GIS analysis there are different techniques that you can apply to the spatial data to solve diverse types of questions. For example, you might want to find the best groundwater wells in a basin that has high yield and good water quality, or the best location to build a treatment plant, etc. Using the correct geoprocessing tool is essential for the success of the analysis.

Geoprocessing refers to the tools and processes used to generate derived data sets from other data using a set of tools. The geoprocessing in GIS is a very important tool in the ArcGIS software and plays a fundamental role in spatial analysis. The geoprocessing is a very broad subject in GIS and has many definitions ranging from process study areas for GIS applications, to existing data to manipulate GIS data, or how you compute with GIS data. Geoprocessing can range from simple tasks to very complicated spatial analyses that aid to address an important spatial problem. A typical geoprocessing operation takes an input dataset, performs an operation on that dataset generating new information returns as an output dataset. Geoprocessing tools perform essential operations on a database, such as projections, conversions, data management, spatial analysis and others.

The most important geoprocessing tools can be divided based on the following tasks:

1. Extracting features
    a. Clip
    b. Erase
    c. Split
2. Combining features
    a. Merge
    b. Append
    c. Dissolve
    d. Buffer
3. Combining geometries and attributes
    a. Union
    b. Intersect
    c. Spatial Join

Accessing geoprocessing in ArcGIS can be through toolbox, ArcObjects, Command line, Scripts, and ModelBuilder. The ArcToolbox includes the Analysis Tools that have a range of operations such as Extract, Overlay, Proximity and Statistics. For example the overlay and proximity answers basic questions in spatial analysis: "What's on top of what?" and "What's

Electronic Supplementary Material: The online version of this chapter (https://doi.org/10.1007/978-3-319-61158-7_9) contains supplementary material, which is available to authorized users.

W. Bajjali, *ArcGIS for Environmental and Water Issues*, Springer Textbooks in Earth Sciences, Geography and Environment, https://doi.org/10.1007/978-3-319-61158-7_9

near what?" The first set of tools is discussed in Overlay analysis and the second set is discussed in Proximity analysis. In ArcGIS there are two ways to find any tool:

1. Using the search window
2. Browsing the ArcToolbox

Scenario 1: You are a geologist working for Ministry of Water Resources and your supervisor asked you to use ArcGIS and work with the digital data from Sultanate of Oman to prepare it for conducting spatial analysis. You have been given 4-files: *Watershed.shp, Stream.shp, MayhaCatch.shp*, and *Soil.shp*. In this exercise you are going to perform the following functions:

1. Dissolve
2. Clip
3. Intersect
4. Merge
5. Buffer
6. Select By Location
7. Convert Graph to Feature
8. Erase

GIS Approach

Dissolve

Dissolve can be used when the user wants to aggregate features based on a specified attribute. In this example, you want to dissolve the **Watershed.shp** based on the *Code* field to create a new shapefile layer that will contain only four records (A, B, C, and D). The Dissolve tool will creates 4 code regions by removing the boundaries between codes, and each code will include the total area and average acres.

1. Start ArcMap and integrate Watershed.shp from \\Ch09\Data\Q1 folder
2. Open the attribute table of the Watershed.shp (the Code field consists of 4-variables: A, B, C, and D)
3. Close the attribute table of Watershed.shp.
4. Click the ArcToolbox button on the Standard Toolbar to display ArcToolbox Toolbar.

Set the Geoprocessing Environment
Before you use any tool, you should set the appropriate geoprocessing environment.

The environments you set here are used by all geoprocessing tools, whether you run them in ArcToolbox, the Command Line, within ModelBuilder, or in a script. The dialog organizes the environments under headings, like General Settings.

5. Right-click an empty space in ArcToolbox and click Environments.
6. In the Environment Settings dialog, click Workspace to expand it.
7. Current Workspace: navigate to your \\Ch09\Data\Q1 folder, click Add.
8. Scratch Workspace: navigate to your \\Ch09\Result folder, click Add
9. Click Ok to exit the Environment Settings dialog box

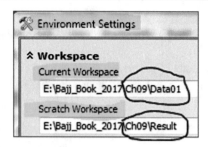

Note: By default, inputs and outputs are directed to your current workspace, but you can redirect the output to another workspace or designate a default scratch workspace for output data.

10. In ArcToolbox, click Data Management Tools/Generalization/Dissolve (or geoprocessing menu/Dissolve)
11. Input features: Watershed.shp
12. Name your output file   *Watershed_Code.shp*
    a. Dissolve_Field(s):   Code
    b. Statistical Field(s):
        i. Area: SUM
        ii. ACRES: MEAN
13. To run the tool click OK

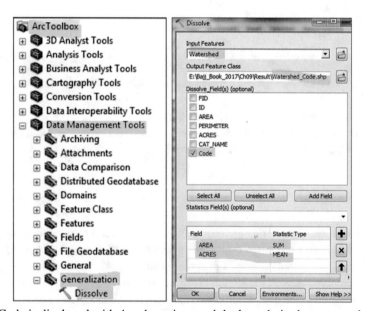

Result: The Watershed_Code is displayed with 4 code regions and the boundaries between codes are removed.

The next step is to classify the Watershed_Code and label each code region with the proper label.

Classification of the New Layer
14. D-click "Watershed_Code"/Symbology/Under the Show click Categories/Unique value/
15. Value Field select the field "Code"
16. Click Add All Values and Uncheck <all other values>
17. OK
18. Open the attribute table of "Watershed_Code" and answer the following questions
    a. How many records the "Watershed_Code" have? Name them
    b. Which region code has the highest and lowest sum area
    c. Which region has the highest average ACRES

Label the New Layer

Label the "Watershed_Code" using the Code field, and make the font Times New Roman and font size 18.

19. R-click "Watershed_Code" and point to Label Features
20. R-click "Watershed_Code" and point to Properties/Labels/Change the Text Symbol to Time New Roman and the size to 18
21. Click OK

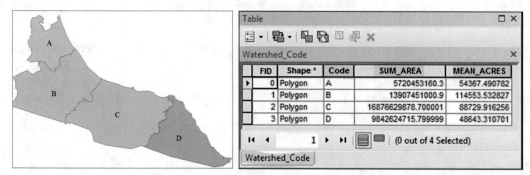

The next step is to create a new layer from the region C, which has the largest sum area of the **Watershed_Code.shp**. The new layer will be called **RegionC.shp**. Then the clip tool will be used to create a new layer from the streams that are located only inside the **RegionC.shp** which will be called "**StreamC.shp**".

22. Click the "Select Feature" button in the Tools toolbar
23. Select region "**C**"

24. In the TOC, r-click **Watershed_Code.shp**/Data/Export Data
25. Browse to \\Result folder and call it **RegionC.shp**, make sure to save as type "Shapefile"
26. Save, then OK, and then Yes to add **RegionC.shp** into the TOC

Clip Tool

The clip tool extracts input features that overlay the clipped features. The clip tool can be used when you want to cut out a piece of one feature class using one or more of the features in another feature class as a "cookie cutter". This is particularly useful for creating a new feature class that contains a geographic subset of the features in another, larger feature class.

Scenario 2: you would like to work with the stream layer that fall inside region C. In this case, you are going to clip the streams using the region C polygon as the cookie cutter to create a new stream feature. The streams will be completely within the region C.

27. Integrate **Stream.shp** from \\Q1 folder
28. Zoom to full extent
29. ArcToolbox/Analysis Tool/Extract/ and D-click Clip (or geoprocessing menu/Clip)
30. Fill the Clip dialog box as below
     a. Input Features: *Stream*
     b. Clip Features: *RegionC*
     c. Output Feature Class: \\Result*StreamC*
31. Click OK
32. Remove from the TOC Stream, Watershed_Code, and Watershed
33. R-click RegionC/Zoom To Layer

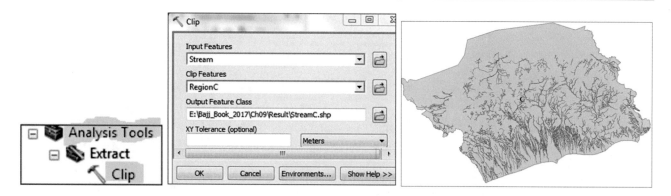

Intersect Tool

The Intersect tool is very similar to the clip tool because the output is defined by the extents of input features. The intersect tool creates a new layer from the common areas of any two selected layers of the same geometry type (such as a polygon). All features that overlap in all layers will be part of the output feature class and the "Intersect" tool preserves the attribute values in both input layers.

Scenario 3: Al Batineh region is the most agricultural area in Oman and your advisor asked you to create the soil map of the Mayha Catchment area from the already existing soil map of north Oman.

34. Insert a new Data Frame and call it **Soil**
35. Integrate MayhaCatch.shp and Soil.shp from \\Ch09\Q1 folder

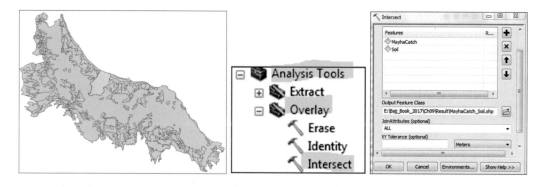

36. ArcToolbox/Overlay/D-click Intersect or (or Geoprocessing menu/Intersect tool)
37. Fill in the intersect dialog box as follows
38. Input Features: MayhaCatch.shp and Soil.shp
39. Output Feature Class: \\Result\Mayha_Soil.shp
40. JoinAttributes: ALL
41. OK
42. Remove all the layers from the TOC, but keep the **Mayha_Soil.shp**
43. R-click **Mayha_Soil.shp**/Zoom To Layer
44. D_click Mayha_Soil/Symbology/Categories/Unique values/Type for the Value Field/Add All Value/Color ramp "Basic random" and Uncheck <all other values>
45. OK

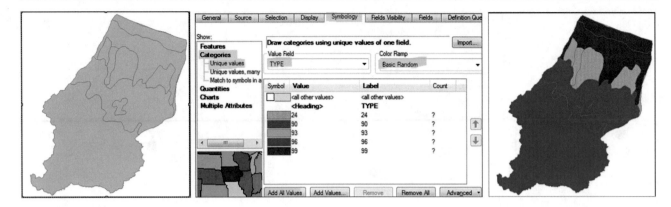

Merge Tool

The Merge tool can be used to combine several layers into a new, single output layer. All input layers must be of the same type. For example, several polygon feature classes can be merged into a polygon feature class. The Merge tool is similar to the Append tool.

   Scenario 4: The agricultural department in the Al Batineh region decided to combine four farms and convert them into one ideal farm. Applying advanced irrigation techniques boost the crop production and reduce the loss of water during irrigation.

46. Insert a new Data Frame and call it Farm
47. Integrate FarmA.shp, FarmB.shp, FarmC.shp, and FarmD.shp from \\Q2 folder
48. ArcToolbox/Data Management Tools/General/Merge or (Geoprocessing menu, Merge tool)
49. Fill in the Merge dialog box as follows
50. Input Datasets:
    a. FarmA
    b. FarmB
    c. FarmC
    d. FarmD
51. Output Dataset: \\Result\FarmNew
52. OK

Result: The four farms become one

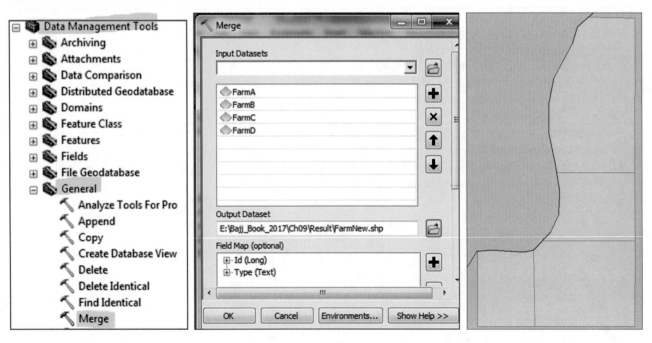

Buffer Tool and Select by Location

Buffer is a zone of a specified distance from a selected feature and it involves the creation of a zone with a specified width around a point, line or area. The result of the buffer is a new polygon, which can be used in queries to determine which entities occur either within or outside the defined buffer zone.

The Select By Location is a special query dialog box that lets you select features based on their location relative to other features.

Scenario 5: Your advisor asked you to identify how many Aflaj "AFLAJ.shp" are located within 500-meters from the outside boundary of the ideal farm (**FarmNew.shp**).These Aflaj will be a source of water for irrigation.

53. ArcToolbox/Analysis Tools/Proximity/Buffer
    - d. Input Features: *FarmNew*
    - e. Output Feature Class: *\\Result\Farm_buffer.shp*
    - f. Linear unit: 500 Meters
    - g. Dissolve Type: All
    - h. Accept the default
54. Ok
55. Place the FarmNew above the Farm_Buffer in the TOC
56. In the TOC, R-click Farm_Buffer, Zoom To Layer
57. Add AFLAJ.shp from \\Q2 folder

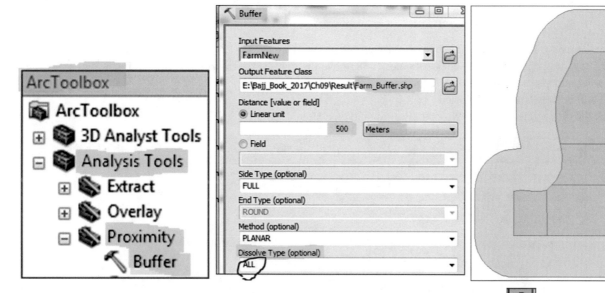

58. Double-click on the symbol under AFLAJ in the TOC/search for Water N/select Water N ![icon]/change the size to 10/ click OK
59. Selection menu/Select By Location
60. Selection method: select features from
61. Target layer(s):   *AFLAJ*
62. Source layer: *Farm_buffer*
63. Spatial selection method for target layer feature(s):   are completely within the Source layer feature
64. Apply and then Close
65. R-click on AFLAJ in the TOC/Data/Export Data save it as **Aflaj_Farm.shp** in \\Result folder
66. OK
67. Click yes to add it in the TOC
68. D-click **Aflaj_Farm** in the TOC/Symbology/Import/select AFLAJ layer
69. OK/OK
70. R-click AFLAJ in the TOC/Remove
71. Double-click on Aflaj_Farm in the TOC to open Properties/click on the Labels tab/check Label features in this layer/Use NUMBER as the Label Field/OK

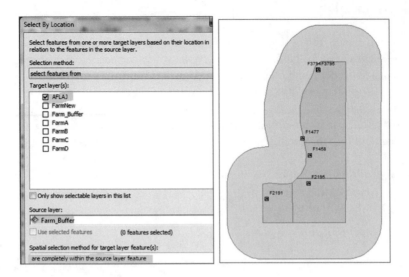

Create an Artificial Water Reservoir for Irrigation Purposes

Scenario 6: You have been to design a reservoir that has a circle shape that could hold at least 40,000 m³ of water. The source of water will come from the Aflaj that are within half a kilometer from the ideal farm. The circle reservoir will have a radius of 50 m and depth of 5.1 m and it will be placed 60 m north of the Aflaj no: 1458. (Volume = 3.14×50 m $\times 50$ m $\times 5.1$ m $= 40,035$ m³)

72. Customize menu, Toolbars, click Draw tool and Zoom in between Aflaj no. 1458 and Aflaj no. 1477
73. In the Draw tool, click the Line , click Aflaj no: 1458, move the mouse straight to the north and D-click (the measurement will be displayed in the lower left corner)
74. A line will be drawn and still selected
75. R-click the drawn line/Properties/Size and Position tab/type Height: 60

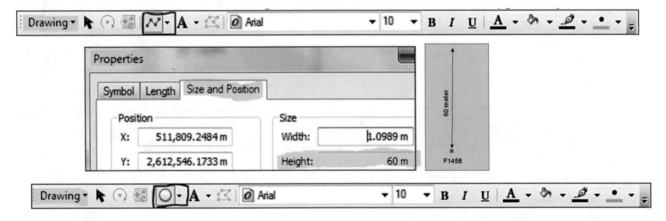

76. In the Draw tool, click the Line drop-down button, click "Circle"
77. Click the upper part of the drawn line to the north of Aflaj no: 1458, and draw a circle with 50 m radius
78. A circle will be drawn and still selected
79. R-click the drawn circle/Properties/Size and Position tab/type Height: 100 m
80. Click Apply/OK Move the circle to the top of the line

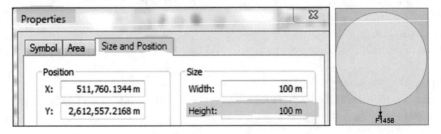

Convert the Graphic into Feature

The circle that is drawn is a graph and you can convert it into a shapefile calling it "**Reservoir.shp**". The Reservoir.shp will be registered in the coordinate of the data frame, which is UTM Z 40 N.

81. In the Draw tool, click the Drawing drop-down button
82. Click Convert Graphics to Features and fill the dialog box as below
    a. Convert: Polygon graphics
    b. Check Selected graphics only
    c. Check "this layer's source data": FarmNew
    d. Output Shapefile: Reservoir.shp
    e. Check Automatically delete graphics after conversion
83. OK
84. Click Yes to add the Reservoir to the TOC

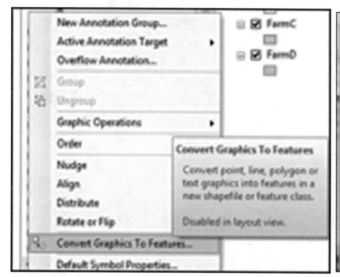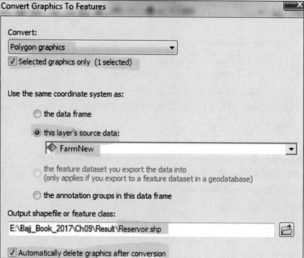

85. Make the Reservoir symbol Hollow and outline color blue and width 1
86. Click OK

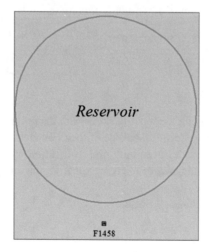

Calculate the Area of the Reservoir
87. Open the Attribute table of the Reservoir layer
88. R-click the Name header/Field Calculator/Yes
89. Under Name = Type "Reservoir" and click OK

90. Click Table Option/Add Field, fill it as below
91. Name: Area
92. Type: Double
93. OK
94. R-click Area/Calculate Geometry/Yes and fill it as below
95. Property: Area
96. Check "Use coordinate system of the data source" PCS: WGS 1972 UTM Zone 40N
97. Units: Square Meters
98. OK

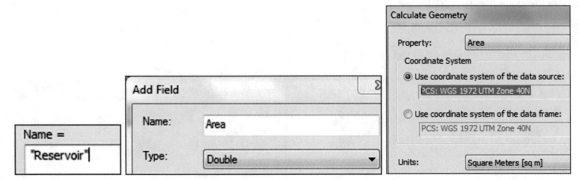

Question: What is the Area of the Reservoir in square meter?
Answer:

Calculate the Area of the Ideal Farm "FarmNew"

 99. Open the Attribute table of the FarmNew
100. Click Table Option/Add Field, fill it as below
101. Name: Area
102. Type: Double
103. OK
104. R-click Area/Calculate Geometry/Yes and fill it as below
105. Property: Area
106. Check "Use coordinate system of the data source" PCS: WGS 1972 UTM Z 40N
107. Units: Square Meters
108. OK

Question: What is the Area of the FarmNew in square meter?

Erase Tool

The Erase tool creates a feature class by overlaying the Input Features with the polygons of the Erase Features. Only those portions of the input features falling outside the erase features, outside of their boundaries are copied to the output feature class.

    Scenario 7: Inside the ideal farm a reservoir has been built and its location will reduce the size of the agricultural area. You are going to use the Erase tool to remove the area of the reservoir from the total farm land.

109. ArcToolbox/Analysis Tool/Overlay/ and D-click Erase
110. Fill the table as below
   a. Input Features: FarmNew
   b. Erase Features: Reservoir
   c. Output Feature Class: \\Result\Farn_Reserv.shp
   d. Click OK

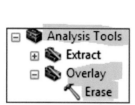
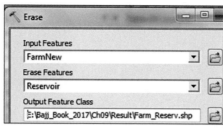

Calculate the New Area of the Ideal Farm "FarmNew" After Removing the Reservoir

111. Open the Attribute table of the Farm_Reserv
112. Click Table Option/Add Field, fill it as below
113. Name: Area_New
114. Type: Double
115. OK
116. R-click Area/Calculate Geometry/Yes and fill it as below
117. Property: Area
118. Check "Use coordinate system of the data source" PCS: WGS 1972 UTM Z 40N
119. Units: Square Meters
120. OK

Question: What is the Area of the **Farm_Reserv** in square meter?

121. Open the attribute table of Farm_Reserv
122. Calculate the total area of the "Area" field and the "Area_New" field

Question: Is the difference between "Area" field and the "Area_New" field equal to the area of the reservoir?

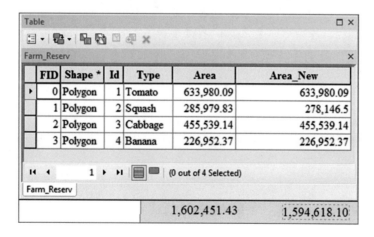

FID	Shape *	Id	Type	Area	Area_New
0	Polygon	1	Tomato	633,980.09	633,980.09
1	Polygon	2	Squash	285,979.83	278,146.5
2	Polygon	3	Cabbage	455,539.14	455,539.14
3	Polygon	4	Banana	226,952.37	226,952.37
				1,602,451.43	1,594,618.10

Site Suitability and Modeling

10

Introduction

Site selection means finding the location that meets a specific condition or criteria. In ArcGIS, you can use different techniques and tools to find the most suitable site for a particular purpose. One of the straightforward procedures that can be used in GIS is modeling. Modeling helps to generalize or simplify an environmental setting and its processes. For example, to find suitable land for a certain purpose such as irrigation or residential building, you have to apply criteria in your analysis to evaluate where the land is most suitable for that particular use. Understanding the main input functions of the suitability modeling guarantees a successful model. The most important elements of the model progression are the following:

1. What is the problem that demands solving?
2. Define a well-articulated criteria for the analysis
3. Gather the necessary data to solve the problem
4. Select the GIS tool needed for the model
5. Create the model to diagram the activity flow

Determine the criteria that allows you to successfully collect the proper data and create the model.

Model 1

The following exercise allows you to use different aspects of functionality in GIS in order to find the most suitable area for building a greenhouse at the Jordan University campus.

Scenario 1: You are a GIS manager at Jordan University and you have been asked by the administrator to choose the best location to build a new Greenhouse on the north-east region of the campus (Fig. 10.1). To build the greenhouse, you have to take in consideration different criteria.

Data: The image was downloaded from Google Earth, and then clipped and georeferenced using the Palestine_1923_ Palestine_Grid projection. The Landuse and Vegetation layers used in this exercise were digitized using the image after it had been rectified.

Electronic Supplementary Material: The online version of this chapter (https://doi.org/10.1007/978-3-319-61158-7_10) contains supplementary material, which is available to authorized users.

Fig. 10.1 Areal image of the
study area

The Criteria to Build the Greenhouse

1. The Greenhouse should be at least 50 m away from the sewer pipeline
2. The Greenhouse should be within a code of 400 of the landuse layer
3. The Greenhouse should be within a Veg_Code 1 or 2 of vegetation layer

GIS Approach

The work will be done using ArcToolbox and ArcMap.
Input files: **Landuse.shp, Pipeline.shp, StudyArea.shp**, and **Vegetation.shp**

1. Start ArcMap and rename the Layer data frame "**Greenhouse**"
2. Connect to Ch10 folder
3. Click Add Data button browse to \\Ch10\Data\M1 folder
4. Highlight LandUse.shp, Pipeline.shp, StudyArea.shp, and Vegetation.shp and click Add
5. In the TOC, drag the StudyArea at the bottom
6. Change the color of the layers
   - Pipeline: Blue, width 2
   - LandUse: Grey 30%
   - Vegetation: Green
   - StudyArea: Hollow

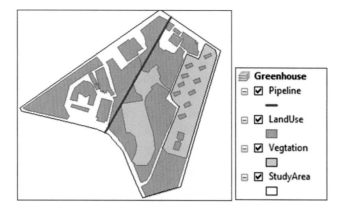

7. Click the ArcToolbox icon in the Standard toolbar
8. Right-click an empty space at the bottom of ArcToolbox/Environment
9. Click Workspace
   - Current Workspace: \\Ch10\M1
   - Scratch Workspace: \\Ch10\Result
10. Click OK

First Criteria: The Greenhouse should be at least 50 m away from the sewer pipeline.

Buffer Analysis: The buffer function will create polygon around the pipeline at our specified distance of 50 m. The buffer shows the area around the pipeline that you can't use for the Greenhouse, as it is too close to the pipeline.

11. ArcToolbox/Analysis Tools/Proximity/
12. D-click Buffer
13. Input Features: Pipeline
14. Output Feature Class: PipeBuffer
15. Linear Unit: 50 m
16. Accept the rest of the default
17. OK

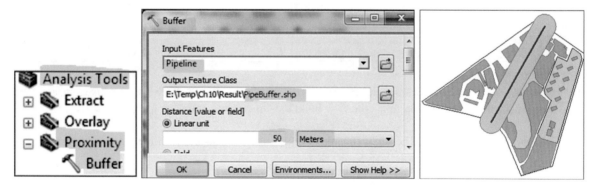

Result: The output buffer is a polygon feature 50 m around the pipeline and the attribute table of the **PipeBuffer.shp**, should have one record.

Overlay Analysis: Union

The Union function is a polygon to polygon overlay method that takes all the features from the input layers, and then calculates the geometric intersection of the layers. The output layer will be of that same geometry type. This means that a number of polygon feature classes and feature layers can be unified together. The output features will have the attributes of all the input features that overlap.

The Union tool will unified the 3-polygon layers: **PipeBuffer**, **Landuse**, and **Vegetation**.

18. ArcToolbox/Analysis Tools/Overlay/D-click on Union tool.

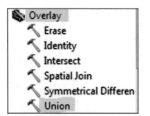

19. Input Features: **PipeBuffer.shp**, **Landuse.shp**, **Vegetation.shp**
20. Output Feature Class: **Union.shp**
21. Accept the default and click OK

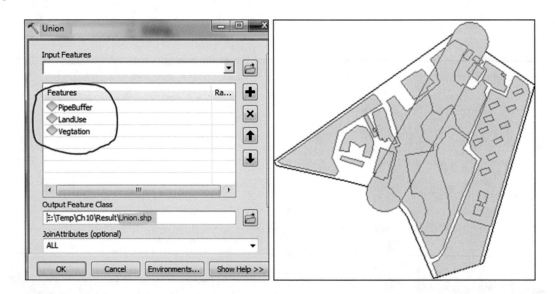

Result: The output file "**Union**" includes all fields from the three input layers: **PipeBuffer.shp**, **Landuse.shp**, and **Vegetation. shp**. The **Union.shp** will be used to select the criteria to build the Greenhouse at the campus.

2nd–3rd Criteria: Greenhouse should be within a code of 400 of the landuse and within a Veg_Code 1 or 2 of vegetation layers.

Select Tool

The select function uses an expression to select features from the input and output layer, which ensures our Structured Query Language (SQL) meets the criteria for the suitability outlined for this assignment.

22. Analysis Tools/Extract/D-click Select
23. Input Features: **Union.shp**

24. Output Feature Class: **Site.shp**
25. SQL statement: "Name" <> 'Sewage Pipe' AND "Veg_Code" <= 2 AND "Code" = 400
26. Click OK/OK.

    Change the color of the Site layer into notable color (e.g. red).

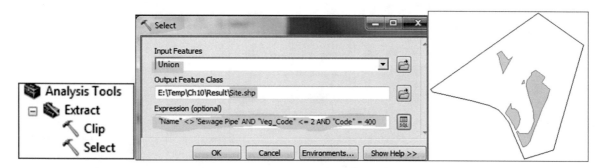

Result: The best potential location selected and consists of two pieces of lands.

Area Calculation

After you've presented the result to the administration, they asked you to calculate the area of the selected locations. The potential location should have a minimum area of 8000 m² as it is going to be used for building the greenhouse.

27. Right-click the Site layer and open the attribute table
28. Open the Table Option/Add Field
29. Name: Area
30. Type: Double
31. Precision: 12
32. Scale: 2
33. Click OK

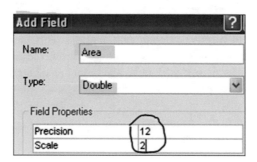

34. Right-click the Area field
35. Calculate geometry/click Yes
36. Make sure the Property is Area and the Units is in Square Meters
37. Accept the other default
38. Click OK

Result: The eastern land will be used as a Greenhouse, as its area is more than 8000 m².

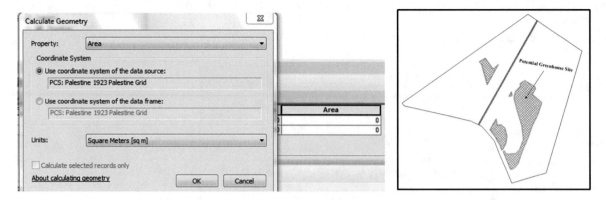

Result: The land that has an area greater than 8000 m² will be the one used for the Greenhouse (as shown in the image).

The area of suitability (pictured in red below) fits all the criteria, which requires the Veg_Code to be equal or less than 2, not within 50 m of the pipeline, and possessing an area of 18,246.01 m² and is in code 400.

Geoprocessing Model for Spatial Analysis

The Geoprocessing Model is a graphical way of systematizing analysis. This means instead of running the tools from ArcToolbox repeatedly, you could automate your analysis as workflow through the geoprocessing model. The model is made up of a process, meaning that each tool is associated with input and output elements. The process consists of four states and each state has the following color:

1. Not ready to run White
2. Ready to run Yellow
3. Running Red
4. Has been run Yellow (drop shadow added to tools and outputs elements)

The model is saved in a created toolbox as a .tbx that can be placed in any folder or the root level of any geodatabase, and the user should have full-write access with it. The model that saved in the custom toolbox becomes a model tool.

ArcGIS allows you to use the ModelBuilder, which is a sequence of tools and data chained together. The output of one tool is fed into the input of another. When you save a model, it becomes a model tool. Models are stored in a user-created toolbox that has full write-access to it. It is recommended to set the environment when working with models, as it sets many parameters such as the output workspace, output spatial reference, and the processing extent.

Model 2

Find Best Suitable Location to Build Nuclear Power Plant

Site selection means finding the location that meets a specific condition or criteria. It is a generalized model that can be used in the GIS environment to find the most suitable site for a particular use. The modeling process identifies the main issue that needs to be answered based on specific criteria. For modeling analysis, the main input function to perform the work is to determine the proper data in order to find the ideal solution. To carry this model in ArcGIS, the exact GIS tool, and the procedures to be carried out, should be determined and understood in advance for an effective result.

Site suitability can be performed by using both raster and vector techniques. This task will be based on applying the vector mode, which relies greatly on proximity and overlay analysis. The main concept will be discerning the sensitive area from the study area and selecting an area that is most suitable.

This exercise allows you do use different aspects of GIS functionality in order to find the most suitable area for building a Nuclear Power Plant (NPP) in Dhuleil Area, Jordan.

The Criteria to Find Suitable Location to Build Nuclear Power Plant

Scenario 2: The Dhuleil area is proposed to be a location to build a Nuclear Power Plant (NPP), as it's an ideal location due to the presence of a plentiful amount of reclaimed water that can be used for cooling purposes. Nevertheless, the main question is: what's the possibility of building a Nuclear Power Plant in the area without affecting the local environment and the water resources?

Dhuleil is an agricultural area which has a major limestone and basalt aquifer. Many wells are tapping these two permeable formations and they have been used for irrigation since the 1960s. In the mid-1980s, a major Sewage Treatment Plant, named Khirbet AlSamra Wastewater Treatment Plant (KSWTP), was built to increase the water resources for irrigation and lessen the use of groundwater whose water quality and quantity had been deteriorated due to extensive use and over exploitation. The area has a network of drainage systems that shifted water from intermittent streams into perennial streams after the KSWTP started to discharge its treated water into the major Zarqa River, which ended up into King Talal Dam reservoir. The water stored into the reservoir was then released in order to irrigate the Jordan Rift Valley, which is one of the most important irrigated areas in Jordan. The surface water and groundwater should be protected from any potential contamination from the proposed NPP.

The Dhuleil area is also highly fractured, and the structure has some influence on the groundwater recharge from precipitation and surface runoff during the wet season. A previous study showed that the wells that are in close proximity to the fault have a slight elevated concentration of nitrate and salinity. Therefore, any NPP should be built at a suitable distance away from the fault system.

The surficial geology in the study area consists of basalt, limestone, alluvium, sandstone, siltstone, and marly limestone. The first four formations are considered highly permeable and are generally associated with moderate to high potential rates of local recharge from rain or surface runoff during wet seasons. The marly limestone and siltstone are considered impermeable surface deposits which have very low to low recharge potential with reduced rates of water movement into or out of these deposits. Therefore, any potential site for building the proposed NNP should be built within these two layers.

The criteria to find best site to build the NPP should be set as follow:

1. The NPP should be at least 300 m away from the main stream
2. The NPP should be at least 200 m away from the faults system
3. The NPP should be at least 500 m away from the groundwater wells
4. The NPP should be built within code 2 (Siltstone & Umm Ghurdan formations)

Building the Geoprocessing ModelBuilder

1. Launch **M2.mxd** from \\Ch10\Data\M2 folder

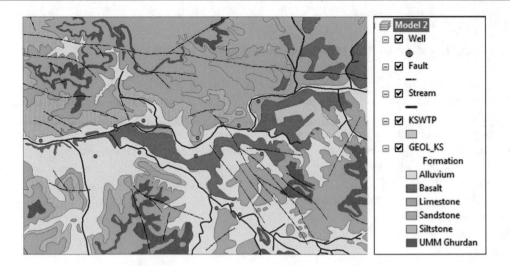

Result: The map open and it has five symbolized layers in the TOC (Well, Fault, Stream, KSWTP, and GEOL_KS).

2. Click Geoprocessing menu/Geoprocessing Options
3. Under General check the "Overwrite the outputs of geoprocessing operations"
4. Click OK

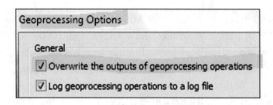

Create a New Toolbox

Create a new Toolbox and store the ModelBuilder inside it. By creating the Toolbox you will have full write to modify it.

5. Click Catalog window
6. Expand Ch10
7. R-click Model\New\Folder and call it Result
8. R-click Result folder\New\Toolbox and rename the new toolbox "**ModelBuilder.tbx**"
9. R-click **ModelBuilder.tbx**/New/Model

Result: The Model window opens in ArcMap.

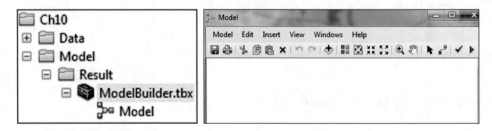

10. Click on the Search window and then type "**Select**".
11. From the displaying research list
12. Drag "Select Layer By Attribute" (Data Management) into the Model.

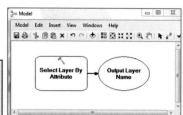

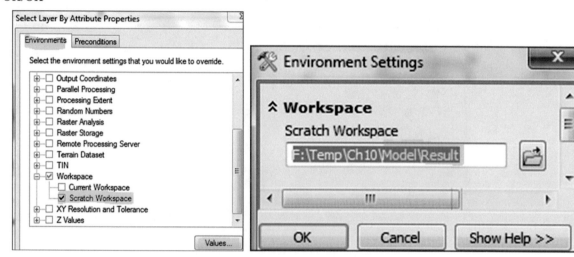

13. Right-click "Select Layer By Attribute" tool in the "Model"
14. Click Properties/click Environments tab/scroll down and Open the Workspace

15. Check Scratch Workspace/click Values/click Workspace to open the Workspace
16. Fill the Scratch Workspace \\Ch10\Model\Result
17. OK/OK

18. D-click "Select Layer By Attribute" tool in the "Model" and fill it with the following.
    • Layer Name or Table View: Stream
    • Selection type (optional): New_Selection
    • SQL: "Status" = 'Main'
19. Click OK.

20. Click Save [icon] icon on the Model

21. Click Full Extent [icon] icon on the Model

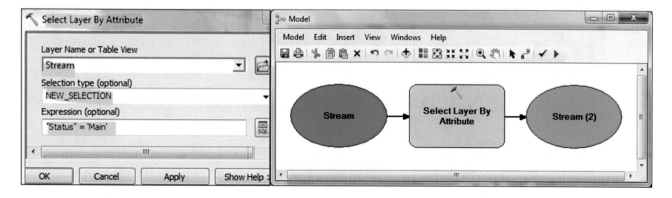

Result: The **process** (the tool and the input and output variables) is in a state **ready-to-run**.

22. Search window/type Buffer
23. Drag Buffer (Analysis) into the Model that you created and place it below Select Layer By Attribute
24. Right-click Buffer tool in the "Model"
25. Properties/Environments tab/scroll down and Open the Workspace
26. Check Scratch Workspace/click Values/open the Workspace
27. Fill the Scratch Workspace \\Ch0\Model\Result
28. OK/OK
29. Double-click Buffer tool in the Model and fill it
    • Input Features: Stream (2)
    • Output Feature Class: \\Result\StreamBuffer.shp
    • Linear unit: 300 m
    • Accept the rest of the default and click OK.

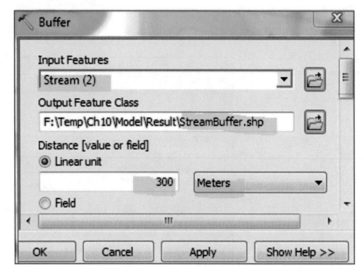

30. Click Save icon on the Model

Result: The **process** is in a state **ready-to-run** and Stream (2) is connected to the Buffer tool.

Click Auto Layout ⊞ icon on the Model to rearrange the two processes and then click Full Extent button.

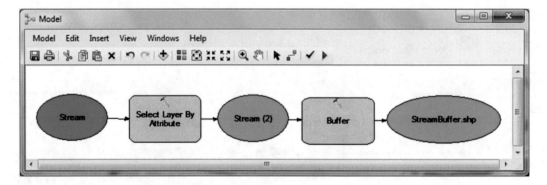

31. Drag Buffer (Analysis) once again into the Model window
32. R-click Buffer (2) tool in the "Model"
33. Properties/Environments tab/scroll down and Open the Workspace
34. Check Scratch Workspace/click Values/open the Workspace

35. Fill the Scratch Workspace \\Ch0\Model\Result
36. OK/OK
37. D-click Buffer (2) tool and fill it
38. Input Features: Fault
39. Output Feature Class: \\Result\FaultBuffer.shp
40. Linear unit: 200 m
41. Accept the rest of the default and click OK
42. OK

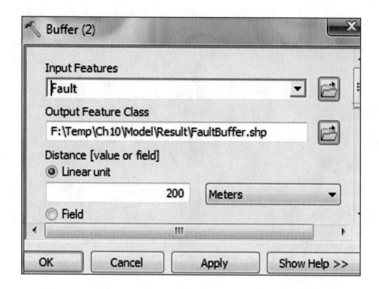

43. Save the Model

Result: The **process** is in a state **ready-to-run**.

44. Click Auto Layout icon to rearrange the three processes and then Full Extent icon

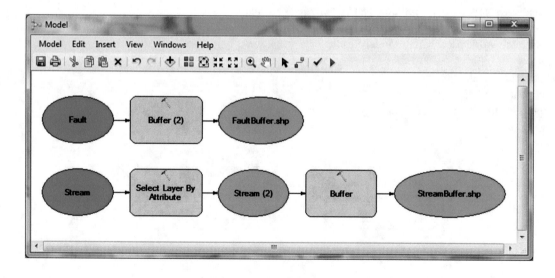

45. Drag Buffer (Analysis) once again into the Model window
46. R-click Buffer (3) tool in the "Model"
47. Properties/Environments tab/scroll down and Open the Workspace
48. Check Scratch Workspace/click Values/open the Workspace

49. Fill the Scratch Workspace \\Ch10\Model\Result
50. OK/OK
51. D-click Buffer (3) tool and fill it
    - Input Features: Well
    - Output Feature Class: \\Result\WellBuffer.shp
    - Linear unit: 500 m
    - Accept the rest of the default and click OK
52. OK
53. Click Auto Layout icon to rearrange the 4-processes and then Full Extent icon

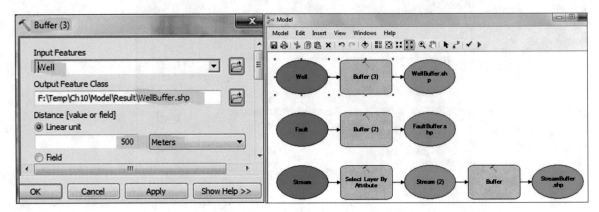

Result: The **process** is in a state **ready-to-run**.

Select Tool

This step is to select the impermeable geological formations that consist of code 2.

54. Search window/Select
55. Drag Select (Analysis) into the Model
56. R-click Select tool in the "Model"
57. Properties/Environments tab/scroll down and Open the Workspace
58. Check Scratch Workspace/click Values/open the Workspace
59. Fill the Scratch Workspace \\Ch10\Model\Result
60. OK/OK
61. D-click Select tool and fill it
    - Input Features: GEOL_KS
    - Output Feature Class: \\Result\GEOL_Code2.shp
    - SQL: "Code" = 2
62. OK/OK

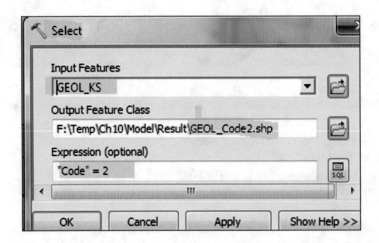

63. Click Auto Layout icon to rearrange all processes and then Full Extent icon

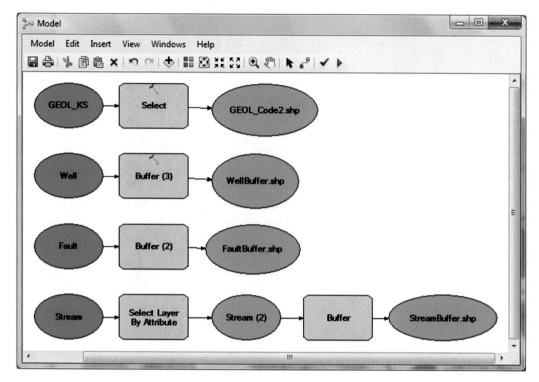

64. Save the Model

Union Tool

Now, you are going to perform a geometric union of the 3-input features (**StreamBuffer.shp**, **FaultBuffer.shp**, and **WellBuffer**). All features and their attributes will be written to the output feature class. The Union tool will only work on polygon feature classes. This step is required in order to combine all the buffers that have been created for the wells, faults, and stream and make them one polygon. Once these features are united, you can then use it in the next step, which will be the erase function.

65. Search window/Union
66. Drag Union (Analysis) into the Model that you created
67. R-click Union tool in the "Model"
68. Properties/Environments tab/scroll down and Open the Workspace
69. Check Scratch Workspace/click Values/open the Workspace
70. Fill the Scratch Workspace \\Ch10\Model\Result
71. OK/OK
72. D-click Union tool and fill it
    • Input Features: StreamBuffer.shp, FaultBuffer.shp, WellBuffer.shp
    • Output Feature Class: \\Result\UnionBuffer.shp
    • Accept the default
73. OK

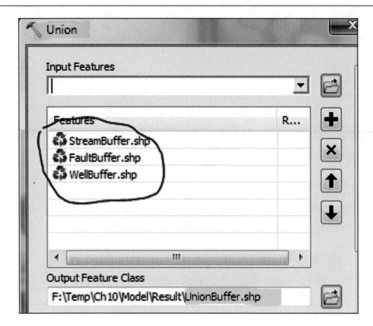

Result: Two things happen.

- The **process** is in a state **ready-to-run**
- The **Union** tool is connected to StreamBuffer.shp, FaultBuffer.shp, and WellBuffer.shp

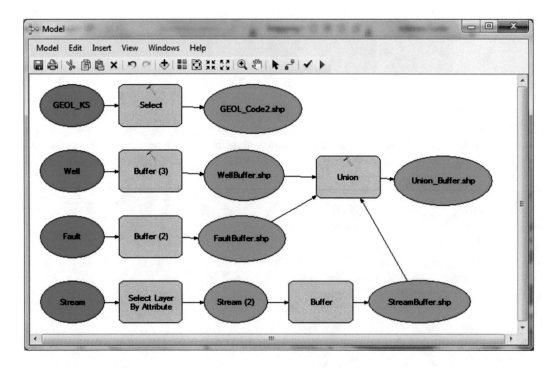

74. Click Auto Layout icon to rearrange all processes and then Full Extent icon

Erase Tool

The Erase tool creates a feature class by overlaying the input features "**UnionBuffer**" that are created from three united layers (**StreamBuffer.shp**, **FaultBuffer.shp**, and **WellBuffer**) with the impermeable geological formation of code 2 (**GEOL_Code2.shp**). Only those portions of the input features that fall outside boundaries of the **UnionBuffer** are copied to the output feature class.

75. Search window/Erase
76. Drag Erase (Analysis) into the Model that you created
77. R-click Erase tool in the "Model"
78. Properties/Environments tab/scroll down and Open the Workspace
79. Check Scratch Workspace/click Values/open the Workspace
80. Fill the Scratch Workspace \\Ch10\Model\Result
81. OK/OK
82. D-click Erase tool and fill it
    - Input Features: GEOL_Code2.shp
    - Erase Features: UnionBuffer.shp
    - Output Feature Class: \\Result\Suitable
83. OK

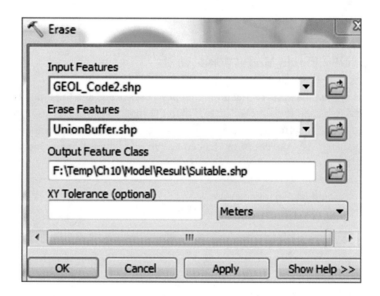

OR you can use the **Connection button** (3rd icon from last)
Use Connection button

84. In the Model window, select the Connect icon
85. Click and Drag a connection from the **GEOL_Code2.shp** to the **Erase** tool
86. From the pop-up menu/select "Input Features"
87. Click and drag a connection line from the **UnionBuffer.shp** to the **Erase** tool
88. From the pop-up menu/select "Erase Features"
89. R-click Suitable.shp/check Add to Display
90. Save the Model window

Result: The Erase process connected and the Model is ready to run.

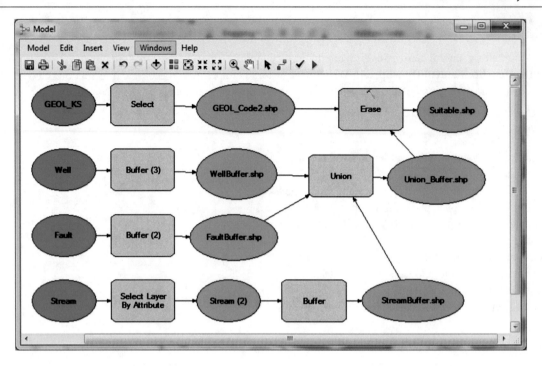

Validate and Run the ModelBuilder in Model Window

When validating the model, the system makes sure that all model processes (tools and the input and output features) are valid to be run properly. If all the variables and tools are validated, then the model will run. If some tools and their variables are not validated, the model will not run. Therefore, if something is wrong, you should open either the variable, or the tool and fix it by providing the correct values. Also, the entire model can be run from the following locations:

1. Model window
2. Model tool dialog box (this should have parameters)
3. Python window
4. In a Script

91. On the Model window, click the Run ▶ icon (last icon)

    Results: Two things happen after running the Model.

- The "Suitable" is added to the table of content.
- Drop shadows will appear around all tools and all output variables. The shadows around the tools and output variables means that the **Model** is in **Has-been-run** state.

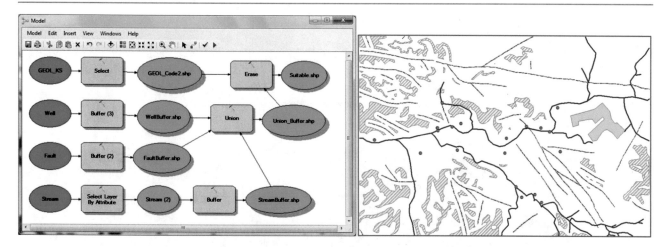

Validate the Model

92. From the Model menu/Click Validate Entire Model

Results: The shadow around the tools and the output variables will be removed.

93. From the Model menu/Click Run Entire Model

Results: The shadow around the tools and the output variables will be restored again

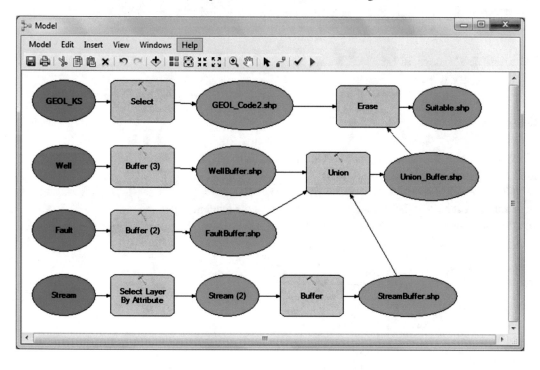

Intermediate Model: In the Model, some of the data is considered intermediate data because once the Model run, the data is no more needed. For example, the "**WellBuffer**" is only created to select the area that's at least 500-meters away from the wells. After that, this variable is no more needed.

94. From the Model menu/Click Delete Intermediate Data

Results: The shadow around the tools (Buffer, Select, Union, and Erase) and all the output variables (with the exception of Stream (2) will be removed. This means the following:

- The Buffer, Select, Union, and Erase are reset because they generate intermediate data.
- The Select Layer By Attribute does not create intermediate data.

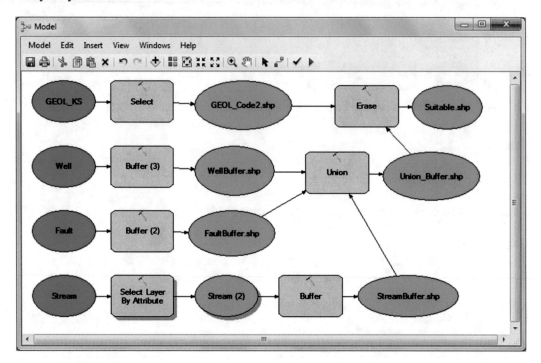

Add Model Name

95. From the Model menu/Click Model Properties/General tab
     a. Name: "**SiteSelection**" (no space is allowed)
     b. Label: NPP Site Selection
     c. Description: Select a site to build a nuclear power plant in the Dhuleil area
     d. Check the Store the relative path name
96. Apply/OK

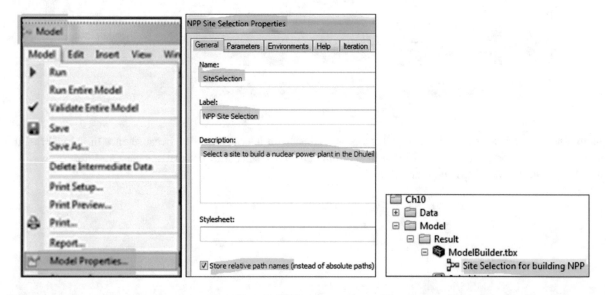

97. Save the Model/then click Model menu and close

Result: The Model will change its name in the Catalog window.

98. Save ArcMap in the \\Result folder and call it NPP.mxd

Set Model Parameters and Run ModelBuilder in Model Tool

The **model tool** is accessed and run through the Catalog window. Prior to run the model, you need to set some model parameters. Till now, the model that you created is a set of connections between the variables (input and output features) and the tools. In the model window, the user can create the needed parameters and these parameters can be used or replaced without changing the model.

99. In the Catalog window/R-click the NPP Site Selection/Open

Result: The "NPP Site Selection" Model tool indicates that it has no parameters.

NOTE: *Do not click the Ok, clicking Ok will run the model*.

100. Click Cancel to close the "**NPP Site Selection**" Model tool
101. In Catalog window R-click "**NPP Site Selection**"/Copy
102. R-click ModelBuilder.tbx/Paste
103. Rename the "**NPP Site Selection (2)**" to Model2

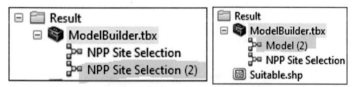

Result: The Model 2 is copied.

100. R-click \\Ch10\Model\New Folder, call it **Result2**
101. R-click Model 2/click Edit to open it in Model window
102. R-click the variable "Stream (2)" and click Model Parameter

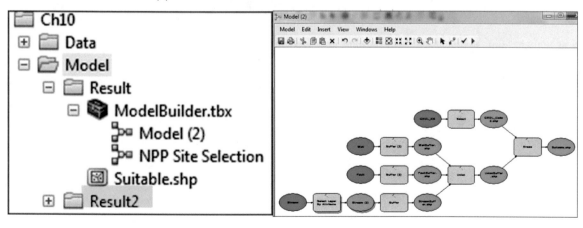

Result: A small letter "P" appear above the element (Stream (2), which indicates that the "**Stream (2)**" element now is designated as an input layer). Therefore, when you are going to run the Model tool in the Catalog window, ArcGIS will ask you to browse for an input layer.

103. R-click the variable "**Stream (2)**" and click Rename and rename it "**MainStream**"/then click OK

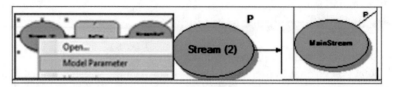

Note: When you ran the buffer tool previously on the stream, you used a 300 meter buffer distance because the distance is an important input function when running the buffer. Therefore, you are going to make the distance as a variable for the Buffer tool and set it as a model parameter. So when you run the Buffer tool, ArcGIS will ask you to specify the distance.

104. R-click the Buffer tool/Make Variable/From Parameters/Distance [value or field]

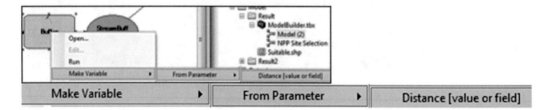

105. Click the Auto Layout ▤▤ icon
106. R-click the variable "**Distance [value or field]**" and click **Model Parameter**

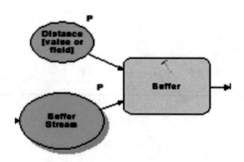

Result: A small letter "P" appear above the "**Distance [value or field]**".

Note: Now you are going to repeat the previous steps on the buffering of the faults and wells.

107. R-click the variable "**Fault**" and click Model Parameter
108. R-click Buffer (2) tool/Make Variable/From Parameters/Distance [value or field]

109. Click the Auto Layout ▤▤ icon
110. R-click the variable "**Distance [value or field] (2)**" and click **Model Parameter**

111. R-click the variable "**Well**" and click Model Parameter
112. R-click Buffer (3) tool/Make Variable/From Parameters/Distance [value or field]

113. Click the Auto Layout icon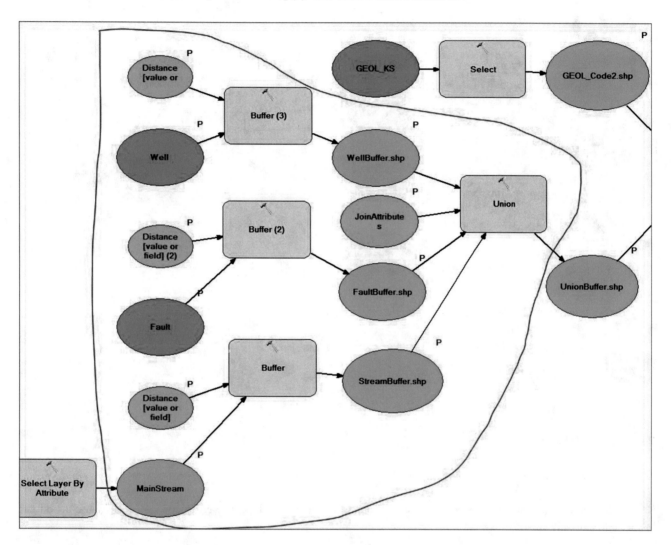
114. R-click the variable "**Distance [value or field] (3)**" and click **Model Parameter**

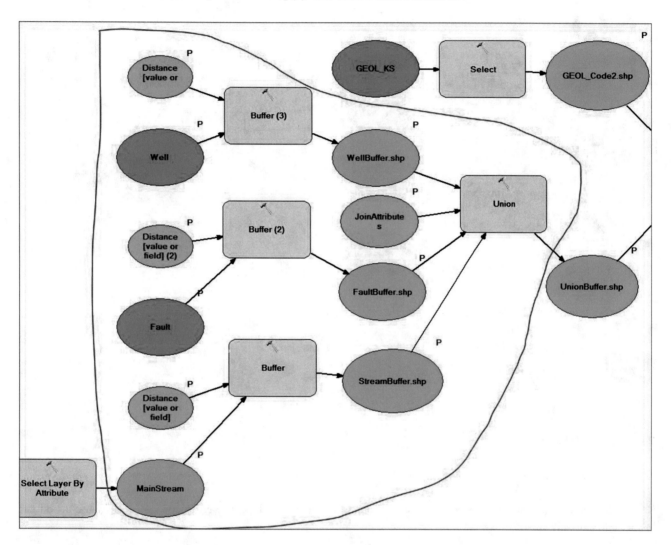

115. R-click the variable "StreamBuffer" and click Model Parameter
116. R-click the variable "FaultBuffer" and click Model Parameter
117. R-click the variable "WellBuffer" and click Model Parameter
118. R-click Union tool/Make Variable/From Parameters/JoinAttributes Click the Auto Layout icon
119. R-click JoinAttributes and click Model Parameter
120. R-click Union_Buffer.shp and click Model Parameter
121. R-click GEOL_Code2.shp and click Model Parameter
122. R-click Suitable and click Model Parameter
123. R-click Suitable/Add to Display

124. Click the Auto Layout icon and then Full Extent
125. Click Model menu and click Validate Entire Model
126. Save the Model and Close

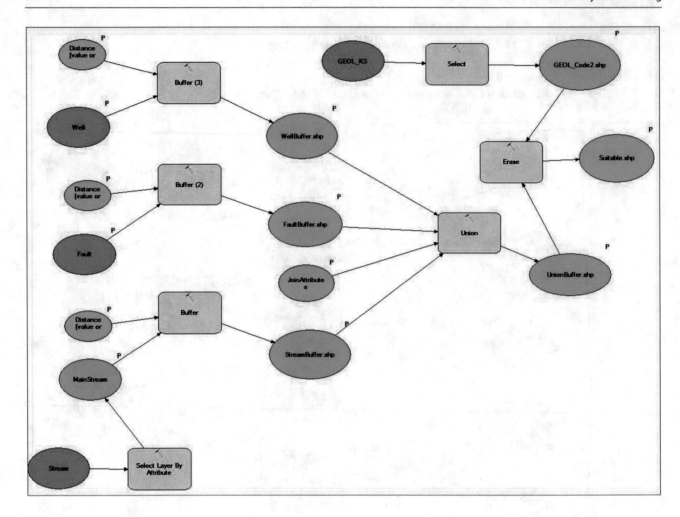

Run the Model Tool

127. In the Catalog window/R-click Model2 and click Open

Result: The Model open and the model parameters appear in the tool dialog box. The error icon appears because the output feature class already exists.

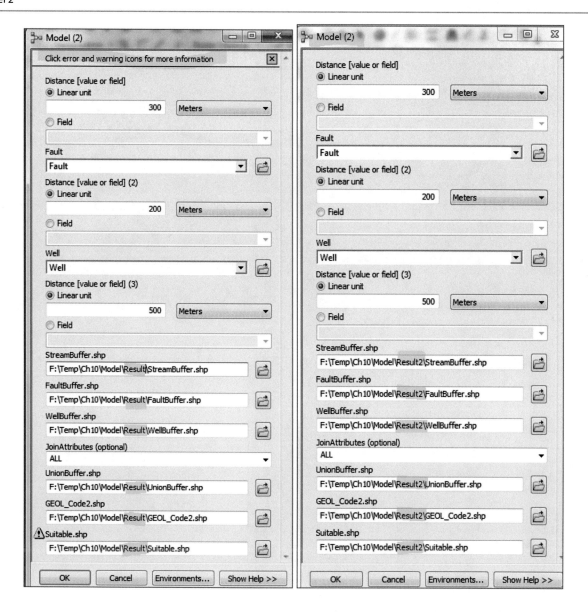

128. Change the path from \\Result into \\Result2
129. Click OK to run the Model tool
130. Click close

Result: The "**Suitable.shp**" and other output layers are added to TOC.

Geocoding

11

Introduction

Geocoding is the process of transforming a description of a location such as an address, or a name of a place to a location on the earth's surface. You can geocode by entering one location description at a time or by providing many of them at once in a table. The resulting output is a location feature with attributes, which can be used for mapping or spatial analysis. With geocoded addresses, you can spatially display the address locations as a points in ArcMap, which help user recognize patterns by using some of the analysis tools available with ArcGIS.

In this chapter, you will learn techniques for finding various types of addresses. The user will be introduced to the preparation of geographic data necessary for address matching called reference data. Technically geocoding is a process of using an address locator to enter address text or a table of addresses to find a corresponding address locations in a geographic database.

Geocoding requires the following:

1. **Address table**: This is a table that includes the addresses that need to be converted into a feature class location.
2. **Reference data**: This is a snapshot of geographic information (point, line, or polygon feature classes) with address information such as streets, or feature class.
3. **Address locator**: An address locator lets you convert textual descriptions of locations into geographic features. Address locators are stored and managed in a workspace you choose. The workspace can be a file folder or geodatabase (file or personal geodatabase).

Once you know what you want to find, the next step to prepare for geocoding is to build or locate sources of geographic data for reference data. In this chapter you are going to perform Geocoding based on

1. Zip Code
2. Street Address

Geocoding Based on Zip Code

In this exercise, you are going to geocode based on the ZIP code. The ZIP code is usually associated with the residents and business addresses. Therefore, utilizing the ZIP code address is an easy approach to be converted into a point feature and using it in ArcMap for analysis.

Scenario 1: About three quarters of the population in Wisconsin and especially those who are living in rural areas rely on groundwater as a source for domestic purposes. Many of these wells are shallow and are susceptible to groundwater contamination, especially from septic tanks on owners' properties of the wells. As a new employer at the USGS, you have been given

Electronic Supplementary Material: The online version of this chapter (https://doi.org/10.1007/978-3-319-61158-7_11) contains supplementary material, which is available to authorized users.

a database table of well owners that their wells have nitrate concentration higher than 20 mg/L. You have been asked to geocode the wells' addresses; based on their zip code and prepare a table showing the average nitrates in the wells in each zip code.

Your duty is to do the following

1. Geocode the database file based on the zip code
2. Summarize the matched addresses based on the zip code
3. Join the summarized matched addresses with the geocoded shapefile

GIS Approach

1. Start ArcMap and rename the Layers data frame "ZIP CODE"
2. Add Data and browse to \\Data\Q1 folder and D-click **GEOCODE.gdb** and highlight **ZipCode_WI** and **Well_Owner** and click Add

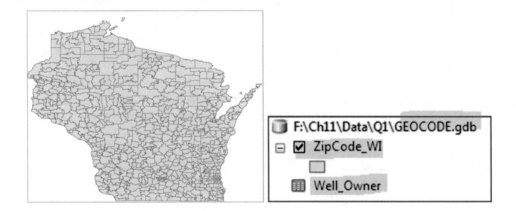

3. In the TOC/r-click **Well_Owner**/Open/ and familiarize yourself with the fields/then close the attribute table

	OBJECTID*	No	Name	Address	City	State	ZipCode	NO3
▶	1	1	Aaron	1614 Gafton Rd.	Madison	WI	53716	40
	2	2	Adam	5022 Ironwood Rd.	Madison	WI	53716	34
	3	3	Adam	1506 Drewry Ln	Madison	WI	53704	35
	4	4	Adam	6834 Park Ridge Dr	Madison	WI	53703	36
	5	5	Adam	3525 Concord Ave	Madison	WI	53714	35
	6	6	Alan	13 Springwood Cir	Madison	WI	53717	41
	7	7	Alan	5706 Odana Rd	Madison	WI	53719	42
	8	8	Alexander	3529 Lexington Ave Apt 3	Madison	WI	53714	62
	9	9	Allen	1329 Temkin Ave Apt 10	Madison	WI	53705	62
	10	10	Angela	141 Metro Ter Unit 101	Madison	WI	53718	62
	11	11	Angela	3309 Brighton Pl	Madison	WI	53713	62
	12	12	Anthony	3507 Milwaukee St Apt A	Madison	WI	53714	62
	13	13	Anthony	505 E Lakeview Ave	Madison	WI	53716	55
	14	14	Antonio	4462 Windsor Rd #4	Madison	WI	53711	63
	15	15	Archie	306 N Brook St Rm 310	Madison	WI	53715	69
	16	16	Attila	4223 Country Road Ab	Madison	WI	53718	72
	17	17	Kvenvolden	3138 Thorp St Apt 7	Madison	WI	53714	46
	18	18	Benjie	3009 Muir Field Rd	Madison	WI	53719	75

Result: The table consists of 779 records and 8 fields. The fields that are relevant to the geocoding are the "Address" and the 5-digit "Zip Code". The table also contains the NO3 concentration.

Create Address Locator

4. Open the Catalog window/Browse to \\Result
5. R-click Result\New\Address Locator
6. In the Create Address Locator dialog box fill it as below
7. Address Locator Style: **General—Single Field**
8. Click OK
9. Reference Data: ZipCode_WI (Make sure Primary Table is selected)
10. *KeyField ZIP (Under Field Map)

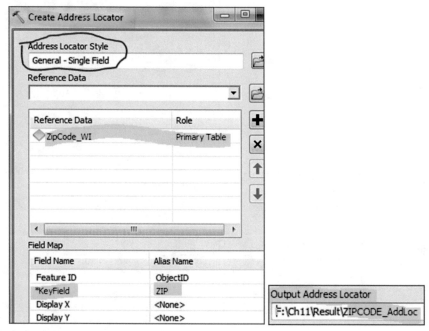

Note: The *KeyField is important to change it to ZIP.

11. Output Address Locator: \\Result\ZIPCODE_AddLoc
12. Save/OK

Result: The address Locator (**ZIPCODE_AddLoc**) is created.

13. Customize menu/Toolbar/Geocoding

14. On the Geocoding toolbar select "**ZIPCODE_AddLoc**"
15. Type 54880 in the <Type an address>/Enter. (This to verify if the "**ZIPCODE_AddLoc**" address locator is established correctly and functioning).

Result: A point will be flashed inside Superior (North-West WI).

Geocode the Addresses

16. Click the Geocode Address button 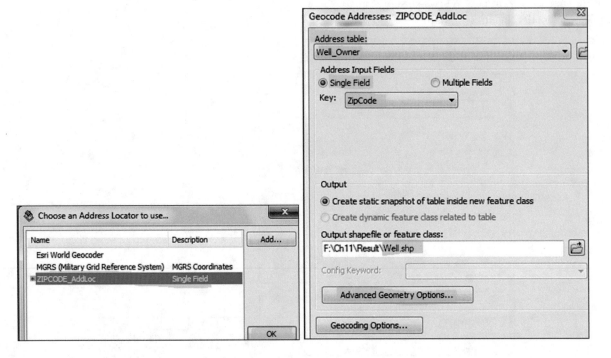 on the Geocoding toolbar
17. Click **ZIPCODE_AddLoc** to select it/then click OK

18. The Geocode Addresses: **ZIPCODE_AddLoc** dialog box display
19. Address table: Well_Owner
20. Check Single Field Key: ZipCode
21. Output shapefile or feature class: \\ch11\Result\Well
22. Click Advanced Geometry Options
23. Make sure the "Use the address Locator's spatial reference" is checked
24. Click OK
25. Click Geocoding Options
26. Under Output field uncheck X and Y coordinates
27. Click OK

Result: Geocoding Addresses dialog box display showing 779 addresses are matched.

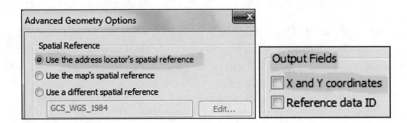

28. Close the Geocoding Addresses dialog box
29. The **Geocoding Result: Well** is now added into the TOC
30. Open the attribute table of the **Geocoding Result: Well**

Result: The **Well** attribute table include fields such as Score, Match_Type, Match_addr, and all the fields from the **Well_ Owner** table that used for geocoding. The Score field means that the candidate are matched or not matched and what is the percent of matching by the address locator. In this scenario all the candidates matched 100%. In the table matching, an address is matched automatically to the candidate with the highest score. The Match_Type field shows how the addresses were matched and in our case it automatically matched (A). The Match_addr field shows where the matched location actually resides based on the information of the matched candidate. For example, an input address of 53202 zip code is matched correctly to a candidate that has the same zip code.

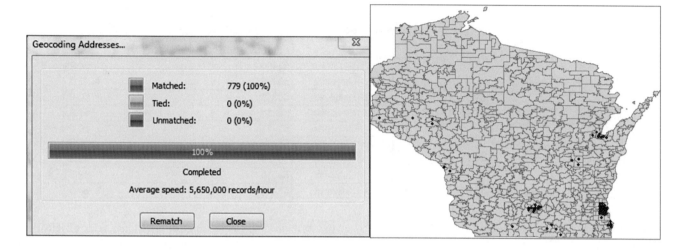

Create Table with the Nitrate Information

This step is important in order to know number of wells in each zip code.

31. Make sure that the attribute table of Geocoding Result: Well is open
32. R-click "Match_adrr" field/Sort Ascending

FID	Shape	Status	Score	Match_type	Match_addr	Addr_type	No	Name	Address	City	State	ZipCode	NO3
563	Point	M	100	A	53132	Address	564	Pamela	8885 S 68th St	Milwauke	WI	53132	47
494	Point	M	100	A	53202	Address	495	Thomas	270 E Highland Ave	Milwauke	WI	53202	53
498	Point	M	100	A	53202	Address	498	Thomas	1209a E Kane Pl	Milwauke	WI	53202	56
510	Point	M	100	A	53202	Address	510	Shaun	1614 N Farwell Ave	Milwauke	WI	53202	29
514	Point	M	100	A	53202	Address	514	Sean	1041 E Knapp St	Milwauke	WI	53202	66
527	Point	M	100	A	53202	Address	529	Robert	1683 N Franklin Pl	Milwauke	WI	53202	42
528	Point	M	100	A	53202.	Address	528	Robert	728 E Juneau Ave	Milwauke	WI	53202	41
532	Point	M	100	A	53202	Address	533	Robert	1108 N Milwaukee St	Milwauke	WI	53202	46
555	Point	M	100	A	53202	Address	556	Phillip	1300 E Kane Pl	Milwauke	WI	53202	63

Note: The "Match_Adrr" field has identical different records as all these wells located in the zone that have the same zip code. Your duty is to know how many wells have the same zip code and their average nitrate.

33. Click the Geoprocessing menu/select Dissolve tool (fill it as below)
34. Input Features: Geocoding Result: Well
35. Output Feature Class: \\Result\Well_Nitrate
36. Dissolve Field: check Match_addr
37. Statistic Field: NO3 and Check MEAN
38. Statistic Field: City and Check COUNT
39. OK
40. Click Yes to add the table to TOC

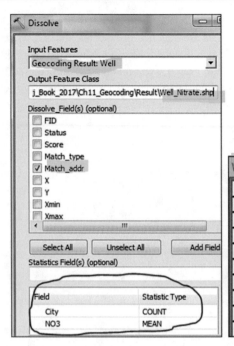

FID	Shape *	Match_addr	COUNT_City	MEAN_NO3
32	Multipoint	53406	10	50.2
8	Multipoint	53209	9	49.8
10	Multipoint	53211	9	50.9
15	Multipoint	53216	9	51.2
41	Multipoint	53711	9	50.4
65	Multipoint	54904	9	50.3
3	Multipoint	53204	8	49.6
9	Multipoint	53210	8	48.5
16	Multipoint	53218	8	54.5

Result: The Well_Nitrate layer display in the TOC and if you open the attribute table you see three fields: Match_addr, COUNT_City, and MEAN_NO3 fields. The COUNT_City field shows how many wells existed in this particular zip code. The MEAN_NO3 is the average of nitrate in all the wells in a particular zip code. For example, the zip code "54904" has 9 wells and their average nitrate is 50.3 mg/L. Some Zip Code, and Zip Addresses include one well and others include many more.

Symbolizing

This step is important to provide symbology for the wells in the zip code. The zip code that has wells with high average nitrate concentration will have bigger symbol, while the zip code that has wells with lower average nitrate concentration will have smaller symbol.

41. D-click Well_Nitrate/Symbology/Quantities/Graduate Symbols
42. Value: MEAN_NO3
43. Click Template/choose Circle 2/Ok/Classes 4 and click Classify
44. Under Break value
45. Type 40, 50, 60, and leave the 70.5
46. Click OK
47. Change the symbol size from 8 to 18
48. Change the color of the symbol as seen below (red, pink, green, and blue)

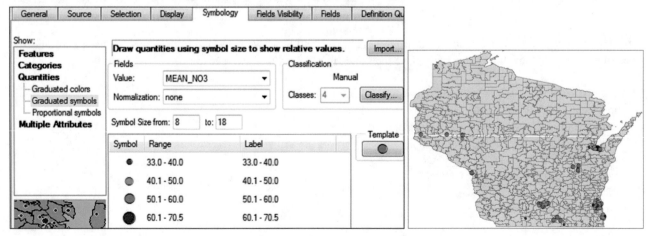

Question: Use Well_Nitrate layer to find out the following:
a) Which Zip code has the highest number of wells and what is the well numbers?
b) Which Zip code has the highest average NO3 concentration and what is the value?
c) Which Zip code has the lowest average NO3 concentration and what is the value?
d) How many Zip code have average NO3 concentration higher than 50 mg/L?
e) In which area of Wisconsin is the highest average NO3?

Geocoding Based on Street Address

Geocoding is a fundamental part of business data management. Every organization maintains address information for each client. In water resources, Wisconsin private wells have an address. The address is stored in a table that contains the well owner's address, well depth, and other relevant information.

Scenario 2: You are working for Douglas County in the city of Superior, you have been given an Excel file that contain well owners and their addresses. You have been asked to geocode the well address owners by converting them into a point shapefile and do some analysis to select the deepest wells that are tapping the sandstone aquifer to be used as an alternative source for water supply in case of emergency.

1. Insert Data Frame call it Superior
2. Add data, browse \\Data\Q2 folder and integrate **Street.shp**
3. Click the symbol of the street and select "**Arterial Street**" from the Symbol Selector

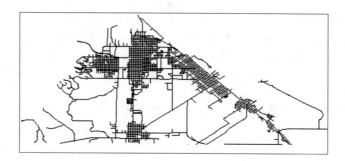

Integrate Excel Table

The excel table **Well.xls** has the information about the wells and your duty is to convert it into a **Table** format in ArcGIS

4. Click Search window/Type Excel/click search icon
5. Click Excel to Table (Conversion) tool
6. Input Excel File: \\Well.xls
7. Output Table: \\Ch11\Result\Table.dbf
8. OK

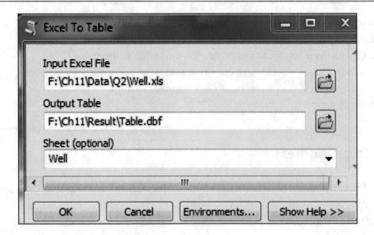

Result: The excel file now is converted into **Table.dbf** and it added into the TOC.

Create Address Locator

Geocoding is performed in ArcGIS with an address locator. An address locator is a dataset stored in either a geodatabase or a file folder that contains information about local conventions for addresses (known as an address locator style) and embedded map data such as street centerlines with address ranges (known as reference data). When you perform geocoding, the address locator interprets an address using the address locator style and finds that address on a map using the reference data. An address locator is created based on a specific locator style. The style determines the type of addresses that can be geocoded, the field mapping for the reference data, and what output information of a match would be returned. It also contains information about how an address is parsed, searching methods for possible matches, and default values for the geocoding options.

9. Click Catalog window/R-click **Result**\New\ Address Locator
10. Address Locator Style: click the browse button, choose **US Address - Dual Ranges/**OK
11. Reference Data: **Street**
12. Role: **Primary Table**
13. Field Map: Filled it as below

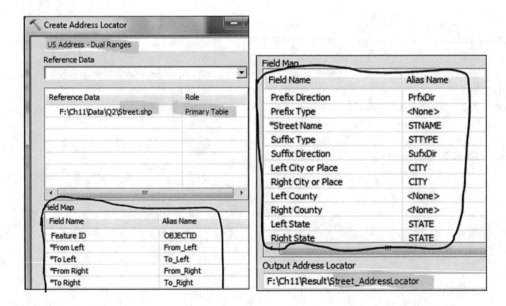

14. Output Address Locator \\Result\Street_AddressLocator
15. OK

Test Your Address Locator

In this step you will test the "**Street_AddressLocator**" that you just created. You will use the Geocoding toolbar to quickly search for an address in city of Superior and display the corresponding location on a map.

16. Make sure that the Geocoding toolbars is available from previous exercise
17. Choose the "**Street_AddressLocator**".
18. In the address box, type **712 N 22ND St, Superior** and press Enter.
19. Right-click the **712 N 22ND St, Superior** in the Geocoding toolbar and click **Add Point**

Result: A graphic point is added to the map.

20. Right-click again the address in the Geocoding toolbar and click Add Callout

Result: A Callout is added to the map.

21. Edit menu/Select All Elements/click Delete button in the keyboard.

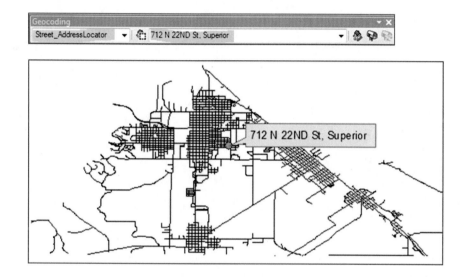

22. In the TOC/R-click Table\Geocode Addresses or click Geocode Addresses 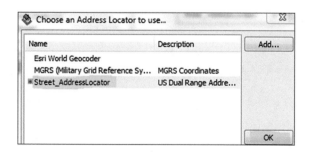 on the Geocoding toolbar (2nd icon)
23. Select "**Street_AddressLocator**"
24. OK

Result: The Geocode Addresses: **Street_AddressLocator** dialog box display.

25. Address table: Table
26. Street or intersection: ADDRESS
27. City or Placement: CITY
28. State: STATE
29. Output shapefile: \\Result\Well_Superior.shp\Save
30. Click Geocoding Options/Fill it as below
31. Spelling Sensitivity: 70
32. Minimum Candidate Score: 70
33. Minimum match score: 70
34. Click OK

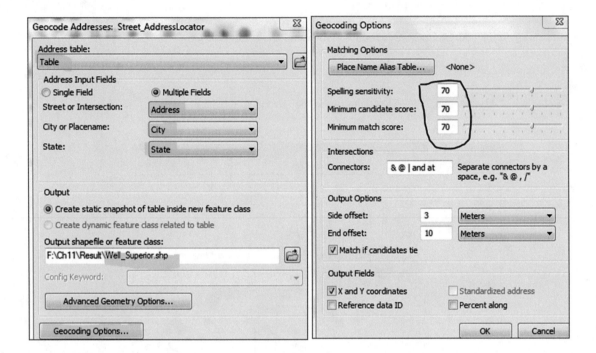

Comment: The spelling sensitivity setting controls how much variation the address locator allows when it searches for likely candidates in the reference data. For example, a low value for spelling sensitivity allows **Univercity** or **Universe** to be treated as match candidates for University. A higher value restricts candidates to exact matches. The Minimum candidate score is when an address locator searches for likely candidates in the reference data, it uses this threshold to filter the results presented. The minimum candidate score for an address locator is a value between 0 and 100. If the address locator seems unable to find any likely candidates for an address that you want to geocode, you can lower this settings so candidates with low scores are presented. The Minimum match score setting lets you control how closely addresses have to match their most likely candidate in the reference data to be considered a match for the address. A perfect match yields a score of 100. A match score between 80 and 99 can generally be considered a good match. An address below the minimum match score is considered to not have a good match.

Click OK to start geocoding

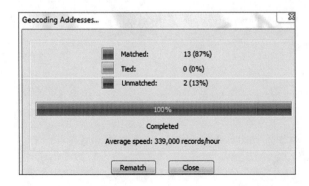

Result: 13 wells matches and 2 unmatched.

35. Click Close

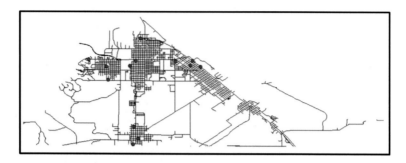

Result: The 13 matched wells are now displayed on the map.

Examine the Geocoding Results

36. Open the attribute table of **Geocoding Result: Well_Superior**

Note: Under the Match_addr field, the unmatched records are empty. The geocoding algorithm was unable to match the two address and now you need to try to match these records. You contacted Douglas County regarding the unmatched two addresses of the wells and they provided you with the correct address (see table below).

Match the Unmatched Addresses

No	Wrong Address	True Address
1	2066 Fisher AV	2022 Fisher AV
4	666 20TH AV	606 20TH AV

37. Close the attribute table

Rematch the Address

38. In TOC, make sure that the **Geocoding Result: Well_Superior** is selected.
39. On the Geocoding toolbar, make sure the Street Locator is selected

40. Click the Review/Rematch Addresses ⬚ button (last icon) on the Geocoding toolbar

41. Maximize the Interactive Rematch dialog box
42. From the **Show results**: choose **Unmatched Addresses**
43. Scroll to the right to view the unmatched addresses

Comment: The Match_addr in the table of content of the "**Well_Superior**" values are blank because the two records are not matched. There may be candidates for a particular record, but there are no candidates with scores of 70 or higher. If there were, they would have been matched because the minimum match score for all your participating locators are 70.

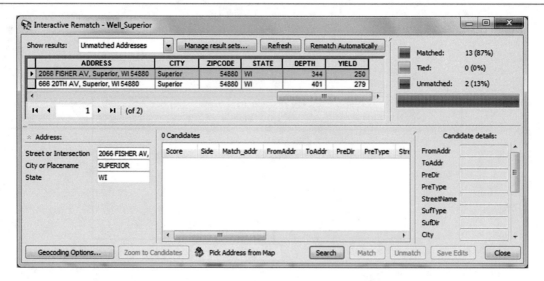

First Address (2066 Fisher AV)

44. Select the first record for which the ADDRESS field has a value of **2066 Fisher AV**
45. Below the Address/Street or Intersection/replace the house number by typing 2022 before the street name Fisher AV, then Enter
46. The candidate score changes to a 100, highlight the 100 score
47. Click Match at the bottom of the Interactive Rematch dialog box

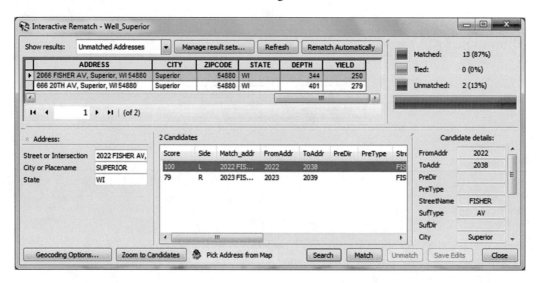

Second Address (666 20TH AV)

48. Select the second record for which the ADDRESS field has a value of 666 20TH AV
49. Change the street number and type 606 before the street name 20TH AV/Enter
50. The candidate score changes to 100.
51. Click Match at the bottom of the Interactive Rematch dialog box
52. Click Close

Result: All the wells now is matched and displayed on the map.

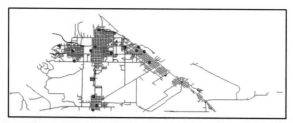

Geocoding Result: Well_Superior

	FID	Shape	Status	Score	Match_type	Match_addr
▶	0	Point	M	100	A	1908 WYOMING AV, Superior, WI
	1	Point	M	100	A	1520 TOWER AV, Superior, WI
	2	Point	M	100	A	6328 BANKS AV, Superior, WI
	3	Point	M	100	M	2022 FISHER AV, Superior, WI
	4	Point	M	92.43	A	1726 N 54TH ST, Superior, WI
	5	Point	M	92.43	M	606 20TH AV E, Superior, WI
	6	Point	M	100	A	302 JOHN AV, Superior, WI

Introduction: Raster Format

The second type of spatial data that is used in GIS is the raster format, which is one form of organization for spatial data. This type of format is suitable for continuous surfaces, such as temperature, elevation, moisture, and much others. Raster is a regular grid of mesh of cells (pixels) that laid over the landscape covering a specific area. The cells of the grid are organized in rows and columns.

Raster can be a satellite imagery, digital aerial photograph, digital picture, scanned map, and save image. The cell is the smallest unit of the raster and each cell has a value representing information, such as elevation. The cell is the fundamental unit of analysis in the raster system, and it represents a location in space. The condition of a given cell is recorded as a numeric value for each cell. The level of detail of features represented by a raster is often dependent on the cell size (spatial resolution).

Resolution means detail with which a map shows the shape and location of geometric feature such as a lake. Smaller cell sizes result in larger raster datasets to represent an entire surface; therefore, there is a need for greater storage space, which often results in longer processing time.

Feature Representation in Raster Format

A point in a vector representation can be approximately transformed to a single cell in a raster representation. Likewise, a vector line can be approximately transformed to a sequence of raster cells lying along that line, and a vector polygon can be approximately transformed to a zone of raster cells overlaying the polygon area. Like vector format, raster provides procedures for deriving new information by transforming or making associations of information from existing layers. GIS analysis using raster data is commonly used in environmental assessments. Raster processes are more commonly used because they can be significantly faster computationally than vector. This chapter consist of four sections and each section include different GIS exercises that deal with raster dataset.

Electronic Supplementary Material: The online version of this chapter (https://doi.org/10.1007/978-3-319-61158-7_12) contains supplementary material, which is available to authorized users.

Section 1: Data Download and Display in ArcMap

1.1. Downloading DEM Image from USGS Webpage
1.2. Explore the DEM image
1.3. Convert image from float to integer

Section 2: Projection and Processing Raster Dataset

2.1. Project an image
2.2. Clip an image
2.3. Merge raster datasets (Mosaic)
2.4. Resample an image
2.5. Classify an image
2.6. Convert Vector Feature into Raster

Section 3: Terrain Analysis

3.1. Create Hillshade
3.2. Create Contour
3.3. Create Vertical Profile
3.4. Create Visibility map
3.5. Create Line of Sight
3.6. Derive Slope and Aspect
3.7. Reclassify the Slope and Aspect
3.8. Combine the Slope and Geology

Section 1: Data Download and Display in ArcMap

Wisconsin Statewide Data Countywide Data	**Douglas County, Wisconsin, United States** Digital Elevation Models (DEM) - 24K Digital Line Graphs (DLG) - 100K

Download DEM Image from USGS Webpage

A popular site from where you can download images (DEM) and integrate directly into ArcGIS is the national map at USGS web page http://nationalmap.gov/. The national map from the USGS allows you to download a digital elevation map (DEM) that can be used directly in GIS.

Download Data from the National Map of USGS

1. Start the internet and in a web browser go to the web page: http://viewer.nationalmap.gov/
2. Under GIS Data title click Download GIS Data

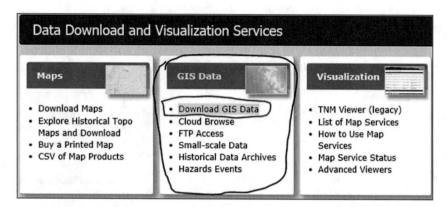

Result: USA map display with a menu and icons.

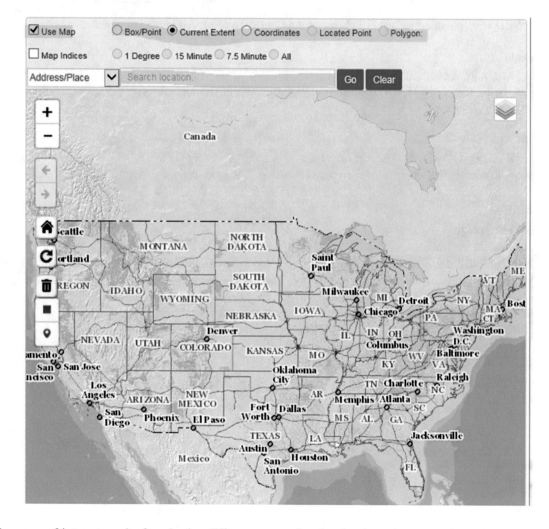

Note: Your area of interest can be found using different approaches, by drawing a box around your area of interest, or by coordinate location, but the easiest way is by search.

3. In the Search window type "Superior" and select Superior, WI, United States
4. The map zoom in the Superior, WI

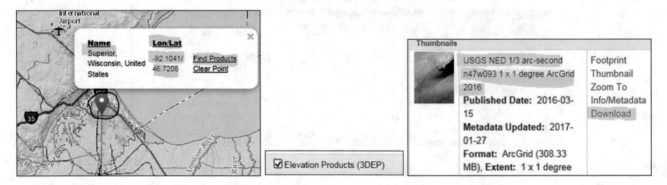

| Superior, Wisconsin, United States | Go | Clear |

5. Click Go tab
6. A dialog box display shows the location of Superior, and its coordinates in Lon/Lat
7. On the left side of the screen, under Data, click the box next to **Elevation Products (3DEP)**
8. Click Find Product

Result: USGS NED 1/3 arc-second n47w093 1x1 degree ArcGrid 2016 displayed in the left panel.

9. Click **Download**
10. Browse and save it in \\Download\NED folder
11. Copy USGS_NED_13_n47w093_ArcGrid.zip from \\Download\NED folder
12. Paste it in \\Data01 folder

Unzip the File
13. R-click USGS_NED_13_n47w093_ArcGrid.zip/Point to WinZip*/Extract to here
*:- Or use any public domain free online unzip program (PeaZip or 7-Zip) to unzip the file

Explore the DEM Image

The "**grdn47w093_13**" DEM is a digital representation of cartographic information in a raster format. DEMs consist of a sampled array of elevations for a number of ground positions at regularly spaced intervals for the city of Superior, WI.

14. Launch ArcMap and call the data frame "**Raster**"
15. Integrate the "**grdn47w093_13**" from \\Data01\ folder (or \\Superior folder)
16. You will be asked to build pyramid
17. Click Yes to create the pyramid
18. The "**grdn47w093_13**" will display

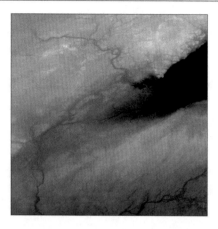

19. In the TOC, R-click the "**grdn47w093_13**"/Properties/click the Source tab
20. The raster property includes four sections that provide detail information about the raster:

First section: Raster Information: This shows that the DEM has 10812 columns and 10812 rows. It shows that the image consists of 1 band and the cell size (resolution) is $9.25.10^{-5} \times 9.25.10^{-5}$ decimal degree. It also indicates that the pixel type is a floating point, meaning that the attribute table of the raster can't be opened in ArcGIS. The pixel depth is 32 bit.

Raster Information	
Columns and Rows	10812, 10812
Number of Bands	1
Cell Size (X, Y)	9.2592593e-005, 9.2592593e-005
Uncompressed Size	445.94 MB
Format	GRID
Source Type	Elevation
Pixel Type	floating point
Pixel Depth	32 Bit

Second section: Extent: This shows the coordinate extent in lat/long.

Third section: Spatial Reference: This shows that the coordinate of the image is registered in GCS_North_American_1983.

Extent	
Top	47.000555555
Left	-93.0005555555
Right	-91.9994444451
Bottom	45.9994444446
Spatial Reference	
XY Coordinate System	GCS_North_American_1983
Linear Unit	
Angular Unit	Degree (0.0174532925199433)

- **Statistics**: This shows the minimum and maximum elevation of the area.

Question: download another image of your interest from http://viewer.nationalmap.gov/. Integrate it in ArcMap and explore it in the source tab. What is the difference between this image and the first image "**grdn47w093_13**" in terms of resolution, format, pixel type, band, and coordinate?

Convert Image from Float to Integer

The values of cells in an integer raster dataset consists of whole numbers and the raster has an attribute table. In a floating point raster, the values of a cell's data consists of numbers with a decimal and the floating raster has no attribute table. Integer values usually represent categorical data, and floating-point values represent continuous surfaces. Note that when converting from floating-point to integer data the following is happening:

1. The value of each cell changes from decimal point to integer (the decimal is simply truncated)
2. The size of the raster dataset will be reduced.
3. If the floating raster values are between 0 and 1, the conversion leads to a loss of information.

The "**grdn47w093_13**" is a floating point and has no attribute table and its cell values is between 176.637 feet (lower elevation) and 490.935 feet (higher elevation).

Int Tool

This exercise will convert the pixel type of "**grdn47w093_13**" from floating point into integer using the "**Int**" tool. The "**Int**" tool will convert each cell value of a raster to an integer by truncation.

21. Insert Data Frame and call it "Integer"
22. Drag the "**grdn47w093_13**" from the **Raster** data frame into the **Integer** data frame
23. Launch ArcToolbox/r-click an empty place below ArcToolbox
24. Click the Environment and click Workspace
25. Current Workspace: \\Ch12\Data01
26. Scratch Workspace: \\Ch12\Result

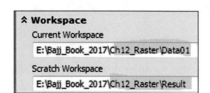

27. OK
28. ArcToolbox/Spatial Analyst Tools/Math/
29. D-click **Int** tool
30. Input Raster: **grdn47w093_13**
31. Output raster: **SupDEM_Int.tif**
32. OK

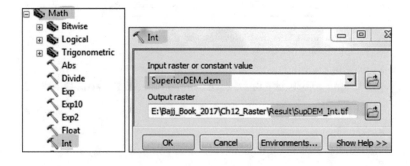

Question: What is the elevation of the "**SupDEM_Int**"?

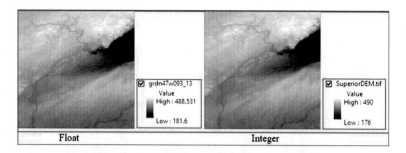

Section 2: Projection and Processing Raster Dataset

Project the DEM of an Area in the Amman-Zarqa Basin

ArcGIS can perform projection of vector and raster. This exercise deals with the international coordinate systems of Jordan. Jordan has specific datum and projected coordinate system. The projection is called the Palestine Projection, which is based on a custom "Transverse Mercator", and the parameters of the projection are listed below:

> False_Easting: 170251.55500000
> False_Northing: 1126867.90900000
> Central_Meridian: 35.21208056
> Scale_Factor: 1.00000000
> Latitude_Of_Origin: 31.73409694
> Linear Unit: meter

In ArcGIS the projection for Jordan is stored in the National Grids/Asia. In this exercise, you are going to project the image of AZ_DEM from GCS_WGS_1984 into Palestine Projection.

1. Insert Data Frame call it "Projection"
2. Integrate **az_dem** from \\Data02 folder
3. The **az_dem** has GCS_WGS_1984
4. Make sure that the ArcToolbox is active
5. ArcToolbox/Data Management Tools/Projections and Transformations\Raster
6. D-click Project Raster
7. Input Raster: az_dem
8. Output Raster Dataset: \\Result\AZ_DEM_PalUTM.tif
9. Output Coordinate System: Click the Spatial reference icon 🗔, open Projected Coordinate Systems\open National Grids, open Asia\select Palestine 1923 Palestine Belt
10. OK
11. Geographic Transformation: select "**Palestine_1923_To_WGS_1984_1**"
12. Accept the default
13. OK
14. Save your map as Projection.mxd in \\Result

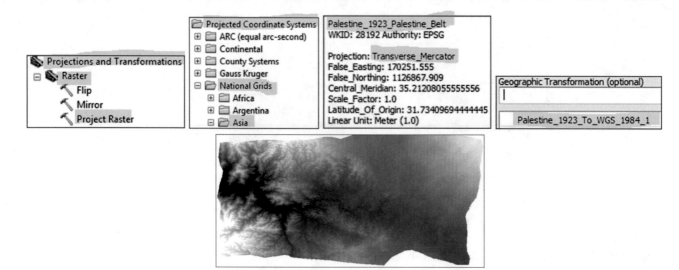

Question 1: What is the resolution of the AZ_DEM_PalUTM.tif?
Question 2: What is the Datum of the AZ_DEM_PalUTM.tif?

Clip an Image

The image (AZ_DEM_PalUTM.tif) that you projected in the previous section is covering a large area of the Amman-Zarqa Basin, and now you are interested in using the Clip Tool to cover only the Dhuleil area. In order to clip the image, you have created a shapefile (**StudyDhuleil.shp**) showing the area that you would like to use in order to clip the image.

1. Insert Data Frame and call it **Clip**
2. Drag **AZ_DEM_PalUTM** from the **Projection** data frame into the **Clip** data frame (or from \\Ch12\Output folder)
3. Add **StudyDhuleil.shp** from \\Data02
4. Click the symbol of the **StudyDhuleil.shp** in the TOC/select Hollow in Symbol Selector
5. Outline Width: 1, and Outline Color: red, click ok
6. ArcToolbox/Data Management Tools\Raster\Raster Processing\Clip
7. Input Raster: **AZ_DEM_PalUTM**
8. Output Extent: **StudyDhuleil**
9. Check Use Input Features for Clipping Geometry (Optional)
10. Output Raster Dataset: \\Result\Dhuleil.tif
11. Make Sure your Clip dialog box is filled as below
12. OK
13. Save your map as Clip.mxd in \\Result

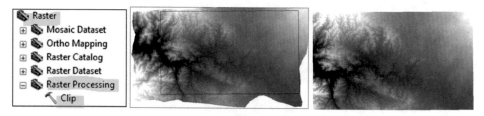

Result: The image is now clipped to the shape of the **StudyDhuleil**.shp.

Merge Raster Datasets (Mosaic)

This tool merges multiple raster dataset into a new raster dataset. To merge the raster datasets, they have to have the same number of bands, same pixel type, and the same pixel depth. When merging the raster dataset in a file format the extension should be specified. There are various extensions that users can choose from and the most popular formats are *.bil* (ESRI BIL), *.bmp* (BMP), *.gif* (GIF), *.png* (PNG), *.tif* (TIFF), *.jpg* (JPEG), *.img* (ERDAS IMAGINE). The extensions will not be added to the name of the raster when raster dataset stored in a geodatabase. In this example we are going to merge **Dhuleil. tif** raster with **KTDam** raster.

14. Insert Data Frame and call it **Mosaic**
15. Integrate into the **Mosaic** data frame **Dhuleil.tif** and **KTDam** raster from \\Data02 folder
16. In the Tools toolbar click Full Extent icon

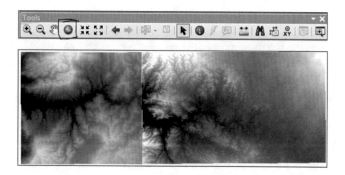

17. R-click the **Dhuleil** raster/Properties/Source and write down the following:
    a. Number of band:
    b. Pixel Type:
    c. Pixel Depth:
18. Repeat the same for the **KTDam** raster
    a. Number of band:
    b. Pixel Type:
    c. Pixel Depth
19. ArcToolbox/Data Management Tools/Raster/Raster Dataset/Mosaic To New Raster
20. Input Rasters: ktdam and Dhuleil.tif
21. Output Location: \\Result
22. Raster Dataset Name with Extension: Dhul_KTDam.tif
23. Spatial Reference for Raster: Click the Spatial reference/Click Add
    coordinate System drop down arrow ⊕ ▾, point to import/browse to \\Data02/select and select KTDam/Add/OK

Note: The coordinate is Palestine_1923_Palestine_Belt.

24. Pixel Type: 16_BIT_SIGNED

25. Number of Bands: 1
26. Accept the rest of the default
27. OK
28. Save your map as Mosaic.mxd in \\Result

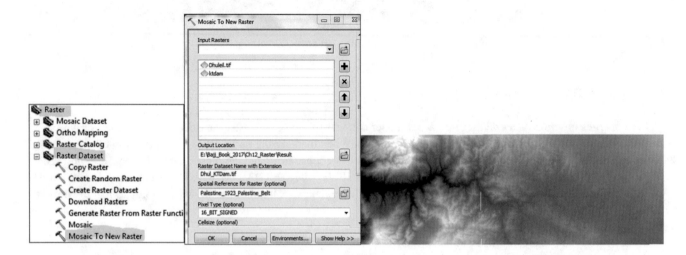

Result: The mosaic raster created from ktdam and Dhuleil.tif as both raster merge together and become one raster.

Resample an Image

Resample tool allows the user to change the resolution of the raster. The cell size can be changed either to a higher or lower resolution, but the extent of the raster dataset will remain the same. The output raster can be saved to any of these formats: BIL, BIP, BMP, BSQ, DAT, Esri Grid, GIF, IMG, JPEG, JPEG 2000, PNG, TIFF, or any geodatabase raster dataset.

There are four options for the Resampling Technique parameter: Nearest, Majority, Bilinear, and Cubic. In this exercise, you are going to use bilinear method and change the resolution (cell size from lower to higher) from 85.72 m into 5 m.

1. Insert Data Frame and call it **Resample**
2. Drag **ktdam** raster from the **Mosaic** data frame into the **Resample** data frame (or \\Data02)
3. R-click **ktdam** raster/Properties/Source
4. Under Raster Information you see that the Cell Size is 85.718 m

Raster Information	
Columns and Rows	377, 338
Number of Bands	1
Cell Size (X, Y)	85.71874745, 85.71874745

5. Close the Layer Properties dialog box
6. ArcToolbox/Data Management Tools/Raster/Raster Processing/Resample
7. Input Raster: **ktdam**
8. Ouput Raster Dataset: \\Result\ktdam_5
9. Output Cell Size: X = 5, Y = 5
10. Resampling Techniques: Bilinear
11. OK

12. R-click **ktdam_5** raster/Properties/Source
13. Under Raster Information you see that the Cell Size is 5 m
14. Save your map as Resample.mxd in \\Result folder

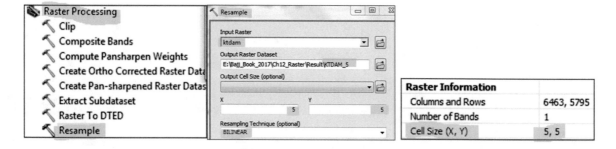

Classify an Image

The classification provides different color to the DEM image

15. D-click **ktdam_5** image/Symbology
16. Make sure the Stretched is selected under the Show:
17. R-click the Color Ramp/Uncheck the Graphic View
18. Scroll down and select Elevation 1
19. OK

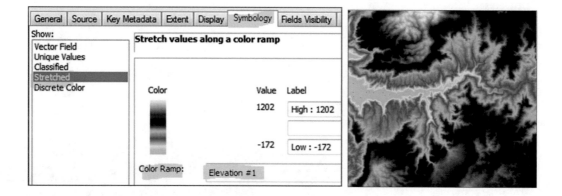

Convert Vector Feature into Raster

The Feature to Raster tool in the Conversion Tools converts features to a raster dataset. Any feature class (geodatabase, shapefile, or coverage) containing point, line, or polygon features can be converted to a raster dataset. This tool is similar to the Polygon to Raster and it always uses the cell center to decide the value of a raster pixel. The tool will be used to convert the geology of Dhuleil area into raster.

20. Insert Data Frame and call it Geology

21. Integrate the Geology.shp from \\Data02 folder
22. D-click Geology layer/Symbology/Categories and select unique value
23. Value Field = Lithology/click Add All Values
24. Ok

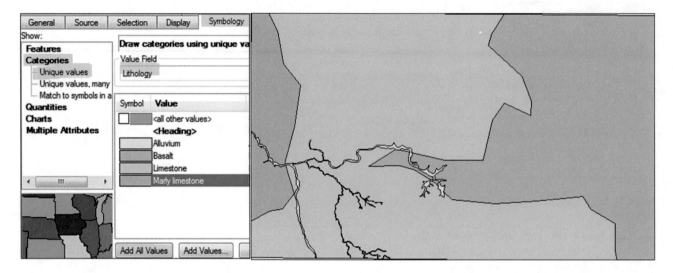

25. ArcToolbox/Conversion Tools/To Raster
26. D-Click Polygon to Raster
27. Input Features: Geology
28. Value field: Lithology
29. Output Raster Dataset: \\Result\Geology.tif
30. Cell assignment type Cell_Center
31. Cellsize: 100
32. OK

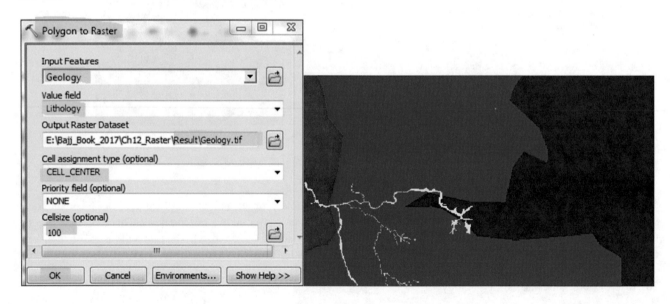

Result: The geology.tif is created with the lithology that consists of 4-geological formations.

Section 3: Terrain Analysis

Create Hillshade

DEM is an excellent raster format that can be used for different terrain analysis. In this exercise you are given a DEM that covers the Dhuleil area. The area is part of the Amman-Zarqa Basin, which is considered one of the most important groundwater basins in Jordan. You have been asked to create a hillshade using the Dhuleil DEM.

The Hillshade tool in ArcGIS creates a shaded relief raster from a DEM raster. The DEM contains all the 3D information about the terrain, but it doesn't look like a 3D object. To get a better expression at the terrain, it is possible to calculate a hillshade, which is a raster format with a 3D-looking image. The Hillshade is a hypothetical illumination of a surface based on a given azimuth and altitude for the sun. It creates a 3-D effect that provides a sense of visual relief for the terrain and is considered the most common way to visualize texture. Using a hillshade enhance the topography of the landscape.

33. Insert Data Frame and call it Hillshade
34. Integrate the DEM of **Dhuleil.tif** from the \\Data02 folder
35. D-click **Dhuleil.tif** image/Symbology
36. Make sure the Stretched is selected under the Show:
37. R-click the Color Ramp/Uncheck the Graphic View
38. Scroll down and select Elevation 1
39. OK
40. Make sure Spatial Analysts are available (Customize/Extensions/check Spatial Analyst)
41. ArcToolbox/Spatial Analyst Tools/Surface/Hillshade
    a. Input Surface: **Dhuleil.tif**
    b. Output raster: \\Result\HillShade.tif
    c. Accept the Default 315 (azimuth), 45 (altitude), and 1 (Z factor)
42. Ok

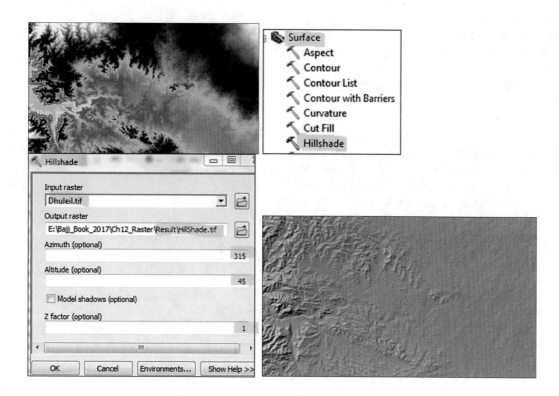

43. R-click Hillshade/Properties/click Display tab
44. Make Transparency (30%)
45. OK

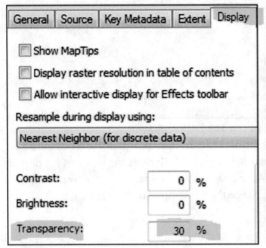

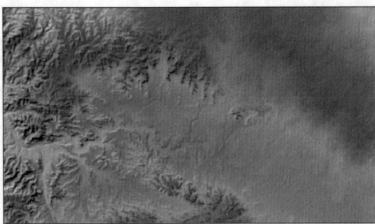

Result: The Transparency applied to the Hillshade raster allows you to see the symbology through the hillshade, yielding a three-dimensional effect.

Create Contour for Dhuleil DEM

The Contour tool creates isolines (contour lines) from the DEM raster. Contour lines are commonly used to represent surface elevations on maps. A contour is a line through all contiguous points with equal height values. The contour will be created for the DEM of Dhuleil.tif, which is registered in UTM coordinate system and the map unit is in meters. The elevation range is 404–905 m above sea level. In this exercise you want to create a 25-m interval from the DEM.

46. Insert Data Frame and call it Contour
47. Drag **Dhuleil.tif** from **Hillshade** data frame into the **Contour** Data Frame
48. ArcToolbox/Spatial Analyst/Surface/and d-click the Contour tool
    a. Input Surface: **Dhuleil.tif**
    b. Output polyline features: **\\Result\Contour25.shp**
    c. Contour Interval: 25 m
    d. Base Contour: 405 m
49. OK
50. D-Click **Contour25.shp** in the TOC/Symbology/Quantities/Graduate Color
51. Set the Value to CONTOUR/4 classes/click Classify/Method select manual
52. Under Break Values type 500, 600, 700, and leave the last value (880)
53. Click an empty place in the Break Value
54. Click OK
55. Click shift and select all contours under Range and Label
56. Click Symbol tab/Properties for Selected Symbols/Width 0.75/OK/OK

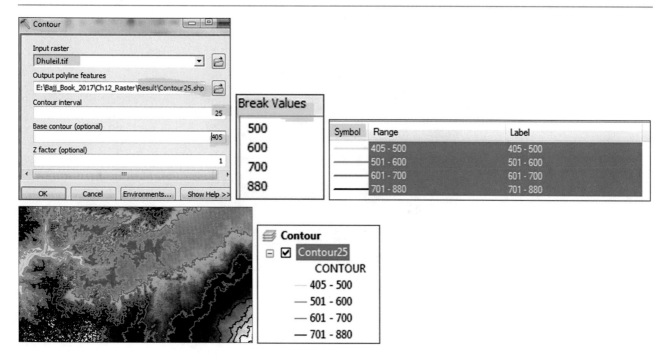

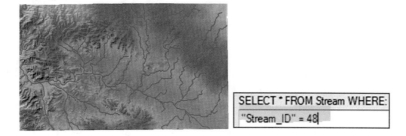

Create Vertical Profile

Profile is a useful GIS operation for terrain analysis, and is a very effective tool for viewing the landscape form. It is created by drawing a line across an elevation image, reading elevation along the line, and then plotting the shape of the terrain. In geology, creating a profile is very useful to understand the form of the land and the outcropping formations, the river morphology in terms of shape and form. In this exercise, you want to create a profile for one stream in the Dhuleil area.

1. Add **stream.shp** to the Contour data frame from \\Data02 folder
2. Uncheck the Countour25
3. Select the river symbol for the **stream.shp**

4. Open Attribute Table of **Stream.shp**/click Table Options/Select By Attribute
5. Use the SQL "Stream_ID" = 48/Apply/Close

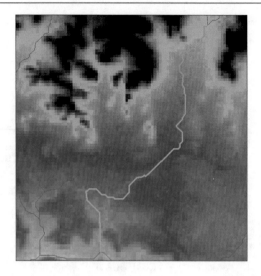

6. R-click stream in TOC/Selection/Zoom to Selected Feature
7. Customize/Toolbar/3-D

8. Click Interpolate Line [icon] on the 3-D toolbar and digitize the selected stream
9. Click Create Profile Graph [icon] tool on the 3-D toolbar

Result: A vertical Profile generated.

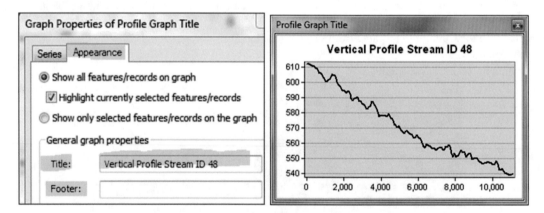

10. R-Click the title bar of the graph
11. Properties/Appearance tab/Title/enter new title" **Vertical Profile Stream ID 48**"
12. Remove the Footer
13. Click OK

Create Visibility Map

The Visibility tool can show what locations in the raster are visible from a specific location and how many observable locations it is visible from. The visibility map is based on two types of analysis:

a) Frequency visibility analysis: determines which raster surface locations are visible to a set of observers
b) Observer visibility analysis: identifies which observers are visible from each raster surface location

14. Insert Data Frame and call it Visibility
15. Drag **Dhuleil.tif** from the Contour data frame into the Visibility data frame
16. Integrate **KSCentroid.shp** from \\Data02 folder
17. ArcToolbox/Spatial Analyst Tool/Surface
18. D-click Visibility
19. Input raster: **Dhuleil.tif**
20. Input point or polyline observer features: KSCentroid
21. Output raster: \\Result\VisibFreq
22. Analysis type: Frequency
23. OK

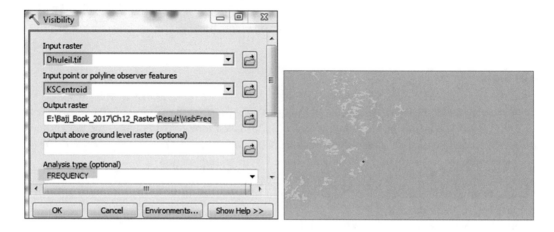

Result: The visibility output map shows the area that can be seen and not seen from the point location KSCentroid.

Add a Height to an Observation Point

24. Open the attribute table of KSCentroid
25. Add field and call it OFFSETA/Short Integer
26. R-click OFFSETA/filed calculator/SQL statement OFFSETA = 25

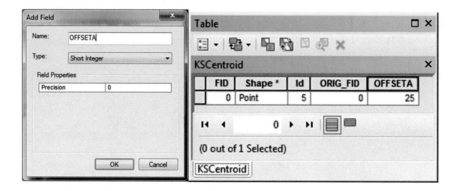

27. R-click KSCentroid/Data/Export Data/Save it as Height25.shp in the \\Result folder
28. Click Yes to add it to TOC
29. Run the visibility map again and call it VisbFreq25 using Dhuleil.tif and Height25.shp
30. OK

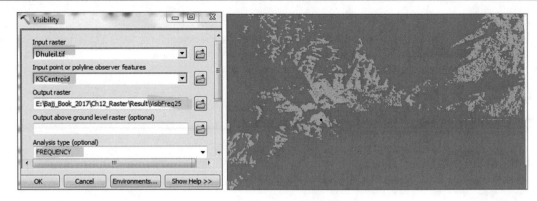

Result: When you added a 25m height the percentage of visible area increases.

Create Line of Sight

Line-of-sight analysis determines whether two points in the raster DEM are visible. To use ths tool a line should be drawn between two location. The Create Line of Sight button on the 3-D Analyst toolbar then divides the drawn line between the two points into segments that are visible (green) from the one point and segments that are invisible (red). The line of sight then can be compared with the visibility map that created in the visibility map section.

31. Integrate **Luhfi_Dam.shp** into the Visibility data frame from **\\Data02** folder
32. Provide the proper symbol to the dam
33. Make sure the 3-D Analyst toolbar is available in the Data View

34. Click Create Line of Sight on the 3-D Analyst toolbar [icon] (third icon)

35. The line of Sight dialog box display
36. Accept the default (observer and Target offset = 0)
37. Click the **Luhfi_Dam** and then click **KSCentroid**

Result: A red color line will be drawn, which mean the dam is not visible from KSCentroid.

38. Use select element [icon] button to select the line of sight and click delete on keyboard

39. Click Create Line of Sight on the 3-D Analyst toolbar again
40. The line of Sight dialog box display
41. Type 10 for the Observer offset and Target offset = 0)
42. Click the **Luhfi_Dam** and then click **KSCentroid**

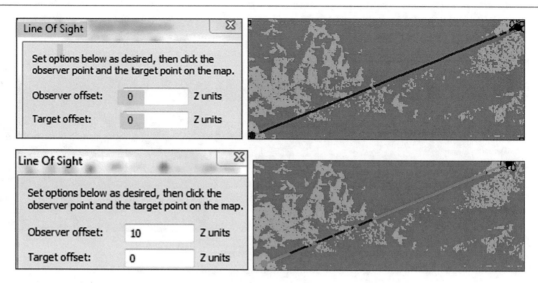

43. A color-coded line will be drawn, green color visible area and red color invisible

Result: The line of sight is correlated nicely with the visibility map.

Slope and Aspect

Slope: the slope is the incline or steepness of the ground surface of a terrain and it can be measured in degrees from horizontal (0–90), or percent slope (rise over run multiplied by 100). For example a slope of 45° equals 100% slope. As slope angle approaches vertical (90°), the percent slope approaches infinity. The slope for a cell in a raster is the steepest slope of a plane defined by the cell and its eight surrounding neighbors.

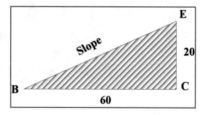

Slope: degree of angel CBE (α)

Slope = (Rise/Run) * 100 = (CE/BC) * 100%

Example: CE = 20, BC = 60; Slope = (20/60) * 100% = 33.3%

Note: If the rise > run, the slope will be more than 100%).

Aspect: The aspect raster map indicates the direction that slopes are facing. The compass direction that a topographic slope faces is usually measured in degrees starting from north. Aspect can be generated from continuous elevation surfaces. For example, the aspect recorded for a TIN face is the steepest downslope direction of the face, and the aspect of a cell in a raster is the steepest downslope direction of a plane defined by the cell and its eight surrounding neighbors.

Derive Slope Layer
44. Insert Data Frame, call it **Slope & Aspect**
45. Drag **Dhuleil.tif** from the **Visibility** data frame
46. ArcToolbox/Spatial Analyst tool/Surface/D-click Slope tool
47. Input raster: Dhuleil.tif

48. Output raster: \\Result\Dhuleil_Slope.tif
49. Output measurement: PERCENT_RISE
50. OK

Q1: What is the range of percent slope values in **Dhuleil_Slope**?

Quiz: Can you run the slope again with "DEGREE" as output measurement?

Classify the Slope into Six Classes

51. D-click **Dhuleil_Slope.tif/**Symbology/Classified (under Show:) 7 Classes/Click Classify/Method choose Manual
52. Under Break Values type 5, 10, 15, 20, 25, 30, and keep the last number 50.52
53. Click empty place below the last number and click OK
54. Click Label/Format label/Number of decimal places = 0/OK/OK

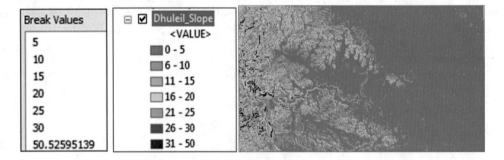

Question: Where is the highest and the gentle slope.

Answer:

Derive Aspect Layer
55. ArcToolbox/Spatial Analyst tool/Surface/Aspect
56. Input raster: Dhuleil.tif
57. Output raster: \\Result\Dhuleil_Aspect
58. OK

Result: Nine principal directions and flat area (−1) are created for the **Dhuleil_Aspect.tif**.

Reclassify Slope and Aspect

The Reclassify tool allows you to change the range of values in the slope and aspect. At the same time it generate an integer raster, which will allows you to see the attribute table and perform further analysis.

Reclassify the Slope
Reclassify the **Dhul_Slope** into six classes

1. ArcToolbox/Spatial Analyst/Reclass/D-click Reclassify
2. Input raster: **Dhuleil_Slope.tif**
3. Reclass Field: **Value**
4. Output raster \\Result**Slope_Reclass.tif**
5. Click **Classify**
6. Classes: **6**
7. Method: **Manual**
8. Under **Break Values** enter **5, 10, 15, 20, 30,** and leave the last value **50.52**
9. OK
10. Click Precision button/Check Use "Format" Precision/Number of Decimal = 0/Ok
11. Check Change missing values to NoData
12. Ok

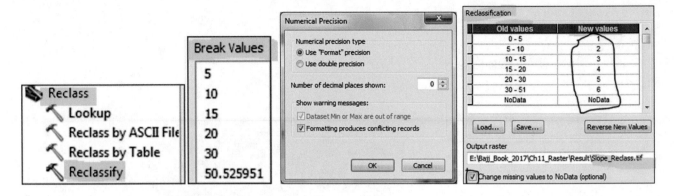

1. Open Attribute Table of **Slope_Reclass.tif**
2. Add Field and call it "Class"/Type Text/Length = 12/click OK
3. Populate the class as below by doing the following
4. Highlight the first row/R-click CLASS field/Field Calculator/Type **"0-5"**/OK
5. Repeat for all the rows as seen in the table below

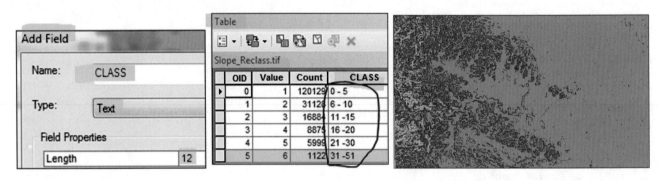

Reclassify the Aspect

Reclassify the **dhuleil_aspect** into 10 classes

1. ArcToolbox/Spatial Analyst/Reclass/Reclassify/
2. Input raster: **Dhuleil_Aspect.tif**
3. Reclass field: **Value**
4. Click **Classify (Classes 10)**
5. Method: **Manual**
6. Under **Break Values** highlight first value and replace it by **−1, then 22.5, 67.5, 112.5, 157.5, 202.5, 247.5, 292.5, 337.5,** and **360**

Note: The difference is 45 with the exception the first class and last class.

7. Click Ok

Note: Now you are going to change the number under the New Values. This can be done by clicking the number and change as in the table below.

8. Click Save \\Result\Aspect_Reclass
9. OK

10. In TOC change the color of the flat area (−1) into white

Old values	New values
1	−1
2	1
3	2
4	3
5	4
6	5
7	6
8	7
9	8
10	1
No data	No data

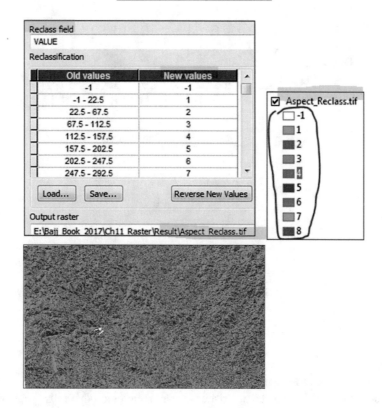

Results:

1. The raster is an integer
2. The −1 is flat

Combine Two Images: Slope and Geology

The Combine tool takes multiple input rasters and assigns a new value for each unique combination of input values in the output raster. The original cell values from each of the inputs is recorded in the attribute table of the output raster. Additional items are added to the output raster's attribute table, one for each input raster.

In the image below, two rasters were input into the Combine function. Notice that each unique combination of values from the two input rasters receives a unique value in the output raster. Two additional fields are added to the output raster attribute table containing the original values from the two input rasters that created the unique combination. Thus, the parentage of the output values can be traced back to the original rasters. Notice that if a cell contains NoData in any of the input rasters, that location will receive NoData for the output. There is no limit to the number of rasters that can be combined; however, there is a practical limit. If there are many rasters all having many different zones, a greater number of unique combinations will be created, resulting in a large attribute table.

Raster01

1	1	0	0
	1	2	2
4	0	0	2
4	0	1	1

Raster02

0	1	1	0
3	3	1	2
	0	0	2
3	2	1	0

Output Raster

1	2	3	4
	5	6	7
	4	4	7
8	9	2	1

Raster 01	Raster 02	Output Raster
1	0	1
1	1	2
0	1	3
0	0	4
1	3	5
2	1	6
2	2	7
0	0	4
0	0	4
2	2	7
4	3	8
0	2	9
1	1	2
1	0	1

Scenario: The Water Authority decided to install a Lysimeter to estimate the infiltration rate in Dhuleil area. The greatest factor controlling infiltration is the amount and characteristics of the precipitation that falls as rain or snow. In general, rain falling on steeply-sloped land runs off more quickly and infiltrates less than water falling on flat land. In addition to that, some outcropping formations allow water to infiltrate at a higher downgradient and recharge the subsurface geology. Alluvium and limestone are more permeable layers that allow portions of the rain to infiltrate into the aquifers and recharge them. Therefore, the aim is to find an area with a slop less than 10% and consists of limestone and alluvium. Slope with value 1 and 2 has a slope less than 10%.

To perform the analysis, the first step is to combine the two rasters: **Slope.tif** and **Geology.tif** together

59. Insert Data Frame call it **Best Location**
60. Integrate **Slope.tif** & **Geology.tif** \\Data03 folder
61. ArcToolbox/Spatial Analyst Tools/Local/D-click Combine
62. Input rasters: **Slope.tif** & **Geology.tif**
63. Output raster: \\Result\Geol_Slope.tif
64. OK

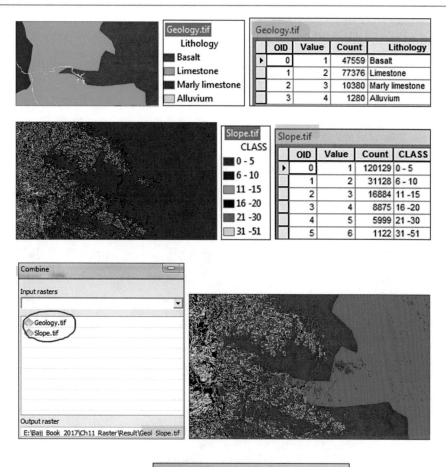

Result: The combine output raster unites the geology and the slope in one map.

To find the areas of limestone and alluvium outcropping formations that have slopes less than 10% slope, you should do spatial analysis and this can be performed using two different approaches: Extract by Attribute or Raster Calculator. Both approaches can be used to find the best location to install the Lysimeter.

Extract by Attribute Tool

This approach allows you to extract the cells of a raster based on a logical query.

65. Spatial Analyst Tools/Extraction/D-click Extract by Attributes
66. In the Extract by Attributes dialog box fill it as follow:
67. Input raster: Geol_Slope.tif
68. Where clause ("Slope" <= 2) AND ("Geology" = 2) OR ("Geology" = 4)
69. Output raster: \\Result\BestLocation.tif
70. OK

Result: The result is BestLocation1.tif that consist of eight records.

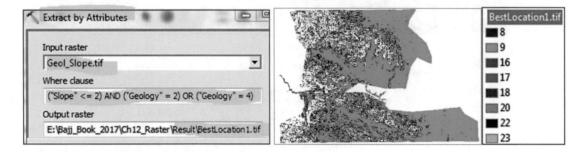

Raster Calculator Tool

In this particular case we have to use the Raster Calculator in the Map Algebra to select the land that has a slope less than 10% and consist of limestone and alluvium using the two rasters: Geology.tif and Slope.tif

71. Spatial Analyst Tools/Map Algebra/D-click Raster Calculator
72. In the Raster Calculator dialog box, type the SQL statement as below

Combine("Slope.tif" <= 2) & ("Geology.tif" == 2) | ("Geology.tif" == 4)

73. Output raster: \\Result\Lys_Location
74. OK

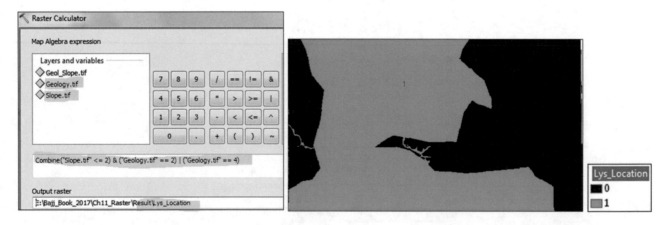

Result: Lys_Location is a raster with two records. Class no. 1 is the location where the lysometer will be installed and the class 0 is the area that are not suitable for lysometer installation.

Calculate the Percent of the Area That Suitable for Installation the Lysimeter

75. Open the attribute table of **Lys_Location** (two records)
76. How many cells value of 1 of the **Lys_Location** raster have?
77. Table Options/Add field/called it "Percentage"/Type = Double) OK
78. R-click field "Percentage" point to Field Calculator
79. Write the following statement: ([COUNT]/(57827 + 78546)) * 100

Result: Value 1 is 57.63%.

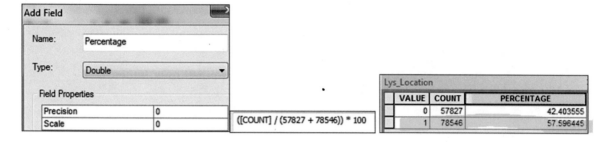

Result: The percentage of area that are suitable for building a Lysimeter is 57.59%.

Introduction

Spatial Interpolation (SI) is a term used to estimate a value of a data variable at an un-sampled site from measurements made in close proximity or within a range of available data. This technique is based on Tobler's First Law of Geography that states that points close together in space are more likely to have similar values than points that are far apart. Use a neighborhood of sample points to estimate a value at an un-sampled location (figure below).

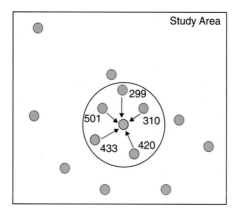

Interpolation uses a neighborhood of sample points of known values (blue color) to estimate a value at an un-sampled location (rose color). This method of estimation is using a specific radius from the un-sampled point.

Various interpolation techniques are used, and these techniques use sample values and X, Y coordinates to estimate the value of an un-sampled point. In general, different methods will generate unalike results with the same input data and no method is more accurate than others under all conditions. Users seeking accuracy should take in consideration a number of point samples and knowledge of the study area.

In order to produce a continuous representation of the phenomenon in question, interpolation makes use of sampling data. There are various methods in GIS that use the interpolation method. Deterministic interpolation techniques create surfaces from measured points. They are based on either the extent of the similarity, an example of such methods is the Inverse Distance Weighted (IDW). There is also a degree of smoothing such as the Trend Surface Analysis method. Geostatistical interpolation techniques such as Kriging are based on statistics and are used for more advanced prediction surface modeling, and also includes errors or uncertainty of predictions. Kriging method is based on the theory of regionalized variables and it is performed by placing an evenly spaced grid over the area for which we have known values and can obtain an estimated surface. The basic idea of Kriging interpolation is that every unknown point can be estimated by the weighted sum of the known points within a certain radius.

Electronic Supplementary Material: The online version of this chapter (https://doi.org/10.1007/978-3-319-61158-7_13) contains supplementary material, which is available to authorized users.

Method of Interpolation

Trend Surface Analysis

Trend surface analysis is a simple way for describing large variations and its function is to find general tendencies of the sample data, rather than to model a surface precisely. The trend analysis calculates the coefficients of a best-fit polynomial surface to fit a set of spatially distributed data points.

In one dimension (1-D): z varies as a linear function of x

$$Z = b_0 + b_1 x + e$$

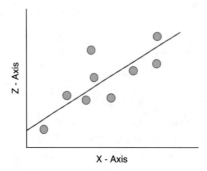

In two dimensions (2-D): z varies as a linear function of x and y

$$Z = b_0 + b_1 x + b_2 y + e$$

where Z is the interpolated parameter.

X and Y are the coordinates of the wells.

b coefficient is estimated from the control points. e: error in prediction.

The aim of this method is to develop a general kind of spatial distribution of an observable fact. The surface can be modeled using a linear or trend surface. Linear trends describe only the major direction and rate of change, while the trend surface provides progressively more complex descriptions of spatial patterns.

Inverse Distance Weighting (IDW)

Inverse distance weighting is a very popular technique in GIS and considers one of the simplest interpolation methods. There are a variety of methods that use weighted moving averages of points within a zone of influence. Interpolation techniques in which interpolation estimates are made based on values at nearby locations weighted only by distance from the interpolation location (figure below).

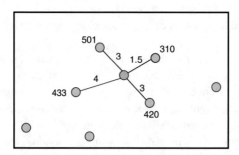

IDW: Closest 4 neighbors

In general the simplified formula for IDW is:

$$V_0 = \frac{\sum_{i=1}^{n} \left(\dfrac{Vi}{Di} \right)}{\sum_{i=1}^{n} \left(\dfrac{1}{Di} \right)}$$

where V_0 is the predictable value at point 0, Vi is the V value at control point i, Di is the distance between control point i and 0, and n is the number of known values used in the evaluation.

The weights are a decreasing function of distance and the user has control over the mathematical form of the weighting function. The size of the neighborhood can be expressed as a radius or a number of points.

Global Polynomial (GP)

Global Polynomial or GP fits a smooth surface that is defined by a polynomial to the input sample points such as the TDS field in the attribute table of the well layer. The GP is similar to taking a piece of paper and fitting it in between the raised TDS values. The result from GP interpolation is a smooth surface that represents gradual trends in the surface over the area of interest. It is used by fitting a surface to the sample points when the surface varies slowly from region to region over the area of interest. While examining and/or removing the effects of long-range or global trends. In such circumstances, the technique is often referred to as trend surface analysis.

Kriging

Using geostatistical techniques, you can create surfaces incorporating the statistical properties of the measured data. Kriging is based on statistics. These techniques produce not only prediction surfaces but also error or uncertainty surfaces, giving you an indication of how good the predictions are. Many kriging methods are associated with geostatistics, but they are all in the kriging family. Ordinary, simple, universal, probability, indicator, and disjunctive kriging, along with their counterparts in cokriging, are all available in the Geostatistical Analyst. Not only do these kriging methods create predictions and error surfaces, they can also produce probability and quantile output maps depending on user needs. Kriging is the estimation procedure using known values and a semi-variogram to determine unknown values. The procedures involved in kriging incorporate measures of error and uncertainty when determine estimations. Based on the semi-variogram used, optimal weights are assigned to unknown values in order to calculate the unknown ones. Since the variogram changes with distance, the weights depend on the known sample distribution. The basic equation used in ordinary kriging is as follows:

$$K(d) = \frac{1}{2n} \sum_{i=1}^{n} \left(Z(xi) - Z(xi + d) \right)^2$$

where d is the distance between known points, n is the number of pairs of sample separated by d; Z is the attribute value (elevation of known points). The equation indicates that the semi-variance is expected to increase as d increases.

One of the most popular approaches is the ordinary kriging, which will be applied in this study. Ordinary kriging assumes the model:

$$Z(\mathbf{s}) = \mu + \varepsilon(\mathbf{s}),$$

where μ is an unknown constant. One of the main issues concerning ordinary kriging is whether the assumption of a constant mean is reasonable. Sometimes there are good scientific reasons to reject this assumption. However, as a simple prediction method, it has remarkable flexibility. The following figure is an example in one spatial dimension:

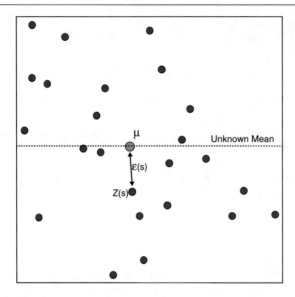

The data is a well with TDS values collected from Maawil watershed in Oman. The wells' locations look like they are distributed randomly. The data is simulated from the ordinary kriging model with a constant mean μ. The true but unknown mean is given by the dashed line. Thus, ordinary kriging can be used for data that seems to have a trend. There is no way to decide, based on the data alone, whether the observed pattern is the result of autocorrelation among the errors $\varepsilon(s)$ with μ constant or trend, with $\mu(s)$ changing with s.

Ordinary kriging can use either semi-variograms or co-variances (which are the mathematical forms you use to express autocorrelation), use transformations and remove trends, and allow for measurement error.

Scenario 1: You are a hydrogeologist working for the water resources in Oman. You have been given a task to evaluate the groundwater along the coast in the Maawil watershed. You have been asked to evaluate, first if the densities of the wells has an effect on the salt intrusion and second, if the quality of groundwater downstream of the two dams, Maawil and Al-Kabir has been improved. The two dams has been built to store the surface runoff produced by rain in the rainy season, and then the stored water in the two dams gradually infiltrate to recharge the aquifer and improve its water quality. To answer these two questions you have been asked to use a GIS technique.

Data and Coordinate System

There are four shapefiles **Dam.shp**, **Stream.shp**, **Watershed.shp** and **Well.shp**. The files are registered in UTM zone 40 and the datum is WGS 1972 (**WGS_1972_UTM_Zone_40N**). Maawil watershed has 1,758 groundwater wells drilled mainly in the upper Maawil catchment area (downstream from the Maawil and Al-Kabir dams). The wells contain complete information about salinity (TDS) and nitrate (NO_3).

Density of Groundwater Well

The Point Density calculates a magnitude-per-unit area from point features that fall within a neighborhood around each cell. Adopting a bigger radii yields a more generalized density raster, and a smaller radius yields a more detailed raster. Only the wells that fall within the neighborhood are considered when calculating the density. If no wells fall within the neighborhood at a particular cell, that cell is assigned no data (NoData).

GIS Approach

1. Start ArcMap and integrate the following file: **Dam.shp**, **Stream.shp**, **Watershed.shp** and **Well.shp** from \\Data\Q1 folder
2. Rename the Layers data frame "Well Density"
3. Symbolize the Dam, Stream, Watershed and Well layers using proper symbols
4. R-click Well.shp/Open Attribute Table

Result: The table contains two fields (TDS and NO_3) that are subject to analysis, close the well table

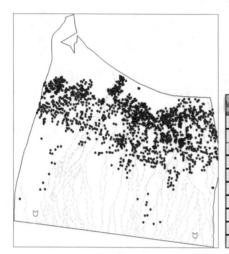

	FID	Well_ID	Shape	WELL_DEPTH	WATER__LEV	YIELD	TDS	NO3
▶	0	1	Point	9.3	8.8		302.7	10.62
	1	2	Point	9.0	8.6		313	15.04
	2	3	Point			4.6	330.9	7.08
	3	4	Point			16.0	345	7.08
	4	5	Point	18.9	17.7	5.0	359.7	4.42
	5	6	Point	17.5	16.1	5.9	367.4	11.06
	6	7	Point	13.8	13.7		369.9	11.95
	7	8	Point	12.9	11.9	0.5	378.7	6.64
	8	9	Point	14.0	13.9		378.9	15.04

Symbolize the Wells Based on TDS

5. Double click Well layer/Symbology/Quantities/Graduate Symbols/Field Value = TDS/Classes 5/click Classify/ Method = Manual/Under Break Values, type 1000, 2000, 3000, 5000, and leave the last number (29632) as it is/click an empty place and click OK
6. Click Label header/Format Labels/Decimal = 0/Check Show thousand separator/OK
7. R-click each symbol of the Well layer in the TOC, change the color based on your taste

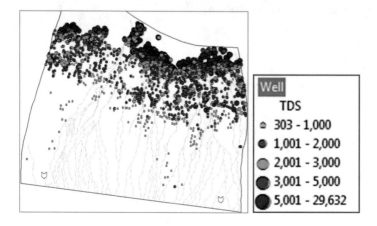

Result: The highest salinity with the bigger symbols is displayed along the coast of Oman.

8. Launch ArcToolbox
9. Right click an empty place below ArcToolbox/Environment/click Workspace
   a. Current Workspace: \\Data\Q1
   b. Scratch Workspace: \\Result
10. Open Raster Analysis in the Environment Setting
    a. Cell Size: As Specified Below
    b. Type: 50
    c. Mask Watershed
11. Click OK

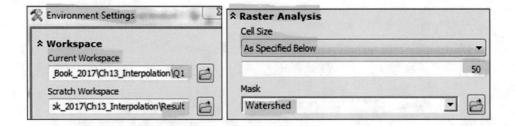

12. ArcToolbox/Spatial Analyst Tools/Density
13. Double click Point Density
14. Input point features: **Well**
15. Population field: **None**
16. Output raster: \\Result\Well_Density.tif
17. Output cell: 50
18. Neighborhood Circle
19. Radius: 50
20. Units: Cell
21. Area Units: SQUARE_KILOMETERS
22. Click OK
23. D-click Well_Density/Symbology/Label Header/Format Labels/Decimal = 0 and show thousands separators/OK/OK
24. Drag Well layer and place it below Well_Density
25. Right click Well_Density/Properties/Display tab/Transparency 30%/OK

Result: The Well_Density raster map displays and shows the cell densities using the salinity (TDS) field. Wells that have high salinity demonstrate bigger densities. This is true that the density of the wells in one location could affect the cone of depression through reducing the water table below the sea level of Gulf of Oman. This causes the sea saline water to invade the shallow aquifer along the coast and increase its salinity to a higher concentration.

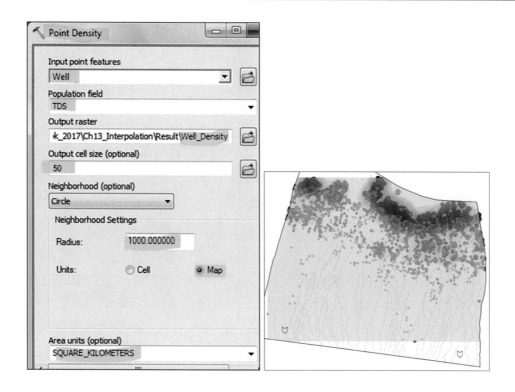

Trend Analysis

The Trend Analysis tool can help identify trends in the input dataset and provides a 3-D perspective of the data. The wells are plotted on the X,Y plane and TDS value is represented by the Z-dimension. Polynomials are then graphed on the scatterplots on the projected planes. An additional feature is that you can rotate the data to isolate directional trends. By default, the tool will select second-order polynomials to show trends in the data, but you may want to investigate polynomials of order one and three to assess how well they fit the data.

The trend surface analysis is a useful tool in early data analysis for delineating basic information and trends regarding the distribution of data. This type of analysis will be performed on the TDS to detect any trend in the salinity

26. Activate the Geostatistical Analyst (Customize/Toolbars)
27. Geostatistical Analyst/Explore Data/Trend Analysis
28. Uncheck Projected Data
29. Uncheck Sticks
30. Layer: Well
31. Attribute: TDS
32. Click the rotate arrow to rotate the graph

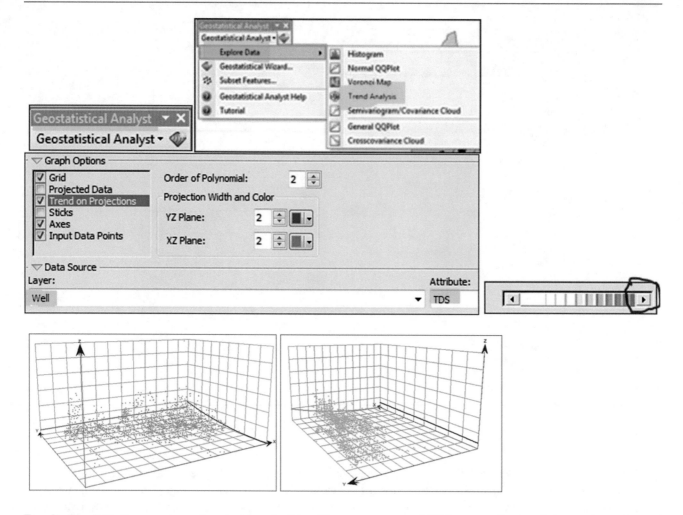

Result: The 3-D diagram shows two trend projections: The west-east trend (XZ – green line) and the north-south trend (YZ-blue). The XZ plane dips from west to east, which means that the TDS concentration increase toward east (toward the coast of Oman) and more or less in the north-south direction. This relationship is clear when you rotate the graph (second image).

33. Close the Trend Analysis

I: Global Polynomial Interpolation

1. Geostatistical Analyst/Geostatistical Wizard
2. Methods
3. Deterministic methods: Global Polynomial Interpolation
4. Input Data/Dataset
5. Source Dataset: **Well**
6. Data Field: **TDS**
7. Next
8. Use Mean

Note: This step allows you to choose the order of polynomial from 1 to 10.

9. Select the **Power** 1
10. Next
11. The Root-Mean-Square (RMS) is **1936.07**
12. Repeat the process by choosing 2, 3, 4, 5, 6, and fill the table below

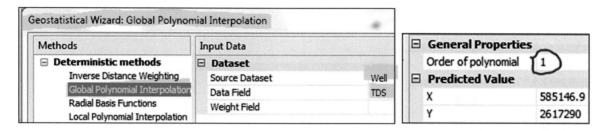

Power	RMS
1	1,936.07
2	1,819.97
3	1,772.62
4	1,727.58
5	1,686.02
6	1,741.25
7	1,777.44
8	**1,472.72**
9	271,845.00
10	20,758.40

Note: This step shows two things: a Scatter plot (predicted values vs measured values) and a Table (Measured, Predicted, and Error).

Result: The best polynomial is when the power is 8 as it generate lower RMS, which is 1472.72.

13. Go back and select the **Power** 8, then click Next
14. Click Finish then click OK

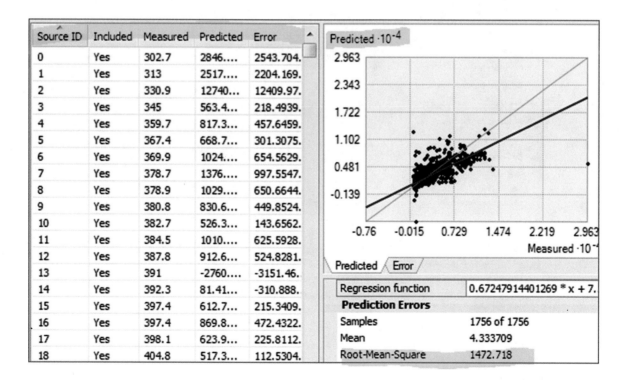

Result: Global Polynomial Interpolation Prediction Map layer is displayed in the TOC and has the same area extent of the Well layer.

Convert the GPI Predicted Map into ESRI Grid and Clip It Using the Mask Technique

Converting the Predicted Map output layer into ESRI grid is an essential step in order to clip the interpolated rater to fit the watershed area using the Mask in the Raster Analysis.

15. Right click GPI Prediction Map/Data/Export to Raster
16. Click the Environment Item and make sure the Workspace and Raster Analysis are set as below:
17. Open Workspace
    a. Current Workspace: \\Data\Q1
    b. Scratch Workspace: \\Result
18. Open Raster Analysis
    a. Cell Size: As Specified Below
    b. Type 50
    c. Mask Watershed
19. OK
20. Input geostatistical layer: Global Polynomial Interpolation
21. Output Raster: \\Result\GPI
22. Output cell size: **50**
23. Click OK

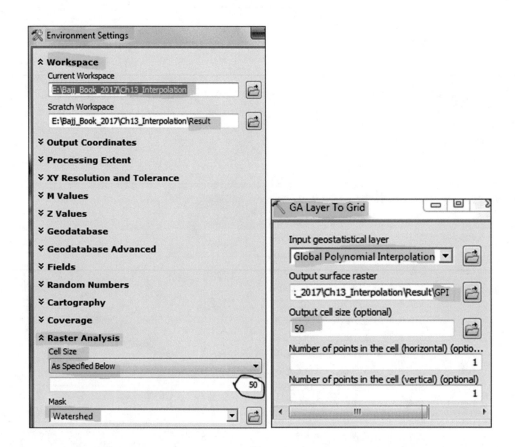

Result: The GPI grid is created and added to the TOC and it is clipped to the Watershed area.

Classify the GPI Map

Classify the grid manually into 10 classes

24. Double click GPI/Symbology tab/under show: highlight Classified/Classes change to 10

25. Click Classify/Method = Manual
26. Under Break Values replace the numbers by 500, 1000, 1500, 2000, 3000, 4000, 5000, 10000, 15000, and 30000
27. Click OK
28. Click Label Header/Format Label/and make the decimal 0
29. Click OK
30. Change the first figure under label to "0–500"/OK
31. In the TOC, r-click the symbol of 15,001–30,000 and change the color into Tuscan Red.
32. Continue changing the color as in the table below

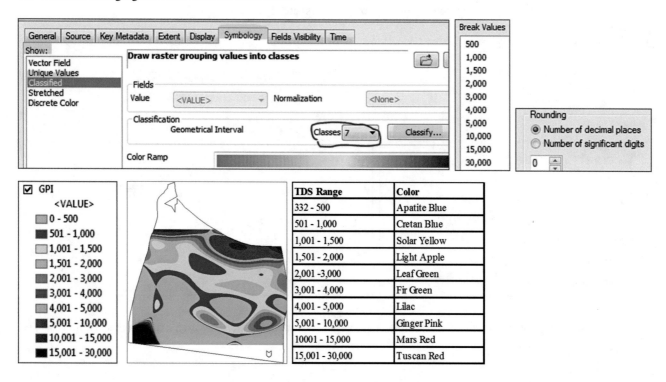

TDS Range	Color
332 - 500	Apatite Blue
501 - 1,000	Cretan Blue
1,001 - 1,500	Solar Yellow
1,501 - 2,000	Light Apple
2,001 -3,000	Leaf Green
3,001 - 4,000	Fir Green
4,001 - 5,000	Lilac
5,001 - 10,000	Ginger Pink
10001 - 15,000	Mars Red
15,001 - 30,000	Tuscan Red

Result: The generated GPI salinity maps for the watershed area indicates that the interpolated surface downstream from the two dams was dominated by the low TDS with the exception from the lower left side of the watershed. In general, the water quality demonstrates improvement away from the coast of Oman.

II: Inverse Distance Weighting (IDW)

You will use now the Inverse Distance Weighting interpolation techniques

1. Insert Data Frame and call it IDW
2. Copy the Well, Stream, Dam, and Watershed layers into it from the GPI data frame
3. Geostatistical Analyst dropdown/Geostaistical Wizard
4. Deterministic methods: Inverse Distance Weighting
5. Source Dataset: Well
6. Data Field: TDS
7. Next
8. Check Use Mean

9. Click to optimize Power value [icon] (icon in the upper right corner)

Result: The power changes from 2 to 1. This means that this is the optimal power value that will generate the minimum RMS, which is 1273.06.

Note: Geostatistical Analyst uses power values greater or equal to 1. When p = 2 (default value), the method is known as the inverse distance squared weighted interpolation. Although there is no theoretical justification to prefer this value over others. The effect of changing "p" should be investigated by previewing the output and examining the cross-validation statistics.

10. Change maximum and minimum neighborhood to 10 and 5 respectively
11. Sector type 4 sectors
12. Click next/Finish/Click OK

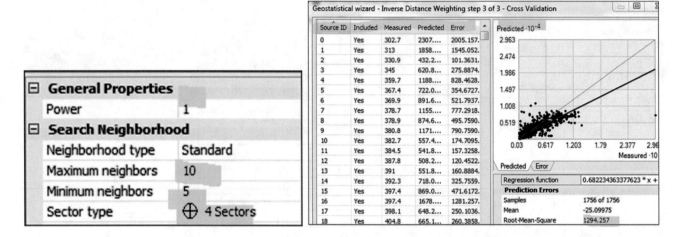

Result: Inverse Distance Weighting Prediction Map is an output layer and has the same area extend as the Well layer. The TDS ranges between 302.7 and 29,632 ppm.

Convert the IDW Predicted Map into ESRI Grid and Clip It Using the Mask Technique

13. Right click IDW Prediction Map/Data/Export to Raster
14. Click the Environment Item and make sure the Workspace and Raster Analysis are set as below:
15. Open Workspace
    a. Current Workspace: \\Data\Q1
    b. Scratch Workspace: \\Result
16. Open Raster Analysis
    c. Cell Size: As Specified Below
    d. Type 50
    e. Mask Watershed
17. Click OK
18. Input geostatistical layer: Inverse Distance Weighting
19. Output Raster: \\Result\IDW
20. Output cell size: **50**
21. Click OK

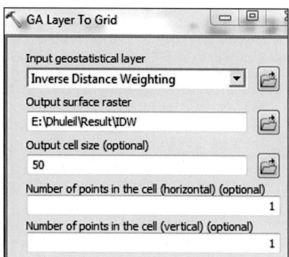

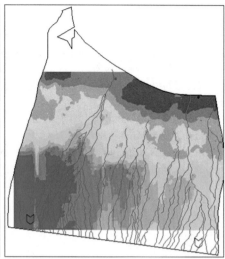

Classify the IDW Map

Classify the **IDW** manually into 10 classes 500, 1000, 1500, 2000, 3000, 4000, 5000, 10000, 15000, and 30000

22. D-click IDW/Symbology tab/under show: highlight Classified/Classes 10
23. Click Classify/Method = Manual
24. Under Break Values replace and the numbers by 500, 1000, 1500, 2000, 3000, 4000, 5000, 10000, 15000, and 30000
25. Click OK
26. Click Label Header/Format Label/and make the decimal 0
27. Click OK
28. In the TOC, r-click the symbol of 15,001–30,000 and change the color into Tuscan Red.
29. Continue changing the color as in the table below

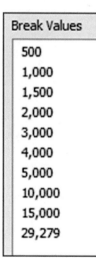

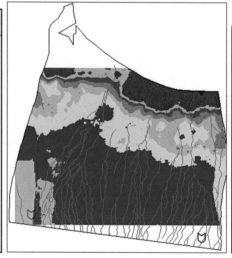

TDS Range	Color
332 - 500	Apatite Blue
501 - 1,000	Cretan Blue
1,001 - 1,500	Solar Yellow
1,501 - 2,000	Light Apple
2,001 -3,000	Leaf Green
3,001 - 4,000	Fir Green
4,001 - 5,000	Lilac
5,001 - 10,000	Ginger Pink
10001 - 15,000	Mars Red
15,001 - 30,000	Tuscan Red

Result: The generated IDW salinity maps for the watershed area indicates that the interpolated surface downstream the two dams was dominated by TDS less than 1,500 mg/l. The salinity of the wells are high along the coast and decrease in the west direction. This shows that the two dams are used as an artificial recharge and are improving the water quality of the aquifer.

Interpolation Using Kriging

Kriging is the best possible interpolation method based on regression against observed z values of surrounding data points and weighted according to spatial covariance values. Kriging assigns weights according to a data-driven weighting function, rather than an arbitrary function. It is still just an interpolation algorithm and will give very similar results to others techniques such as IDW and GPI. There are various types of kriging and in this exercise, you will conduct an interpolation based on Ordinary Kriging.

30. Insert Data Frame and rename it Kriging
31. Copy the Well, Stream, Dam, and Watershed layers into it from the IDW data frame
32. Geostatistical Analyst dropdown/Geostaistical Wizard
33. Geostatistical methods: Kriging/CoKriging
34. Source Dataset: Well
35. Data Field: TDS
36. Next
37. Check Use Mean/OK
38. Kriging Type: Ordinary
39. Output Surface Type: Prediction
40. Next

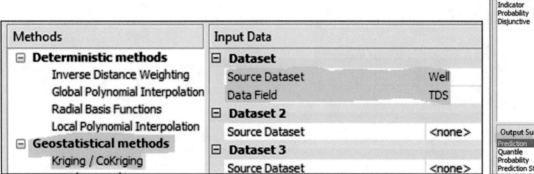

41. Under Model # 1
42. Change the Type to "Gaussian"

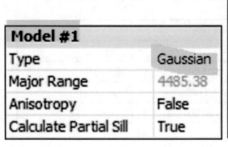

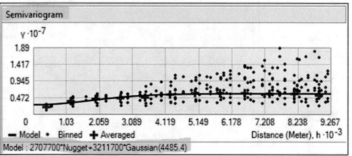

Note: The mathematical model that used in the Semi-variogram allows you to use different models (Stable, Circular, Spherical, Gaussian, Exponential, and others). In this exercise we are going to use the Gaussian model.

43. Under "General"
44. Click the Optimize model then click OK (to calculate the new value for the parameters of Gaussian model (Model # 1) using the iterative cross validation technique)

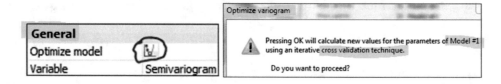

45. Click Next
46. Neighborhood Type: Standard
47. Maximum neighbors: 10
48. Minimum neighbors: 5
49. Sector Type 1 Sector
50. Next

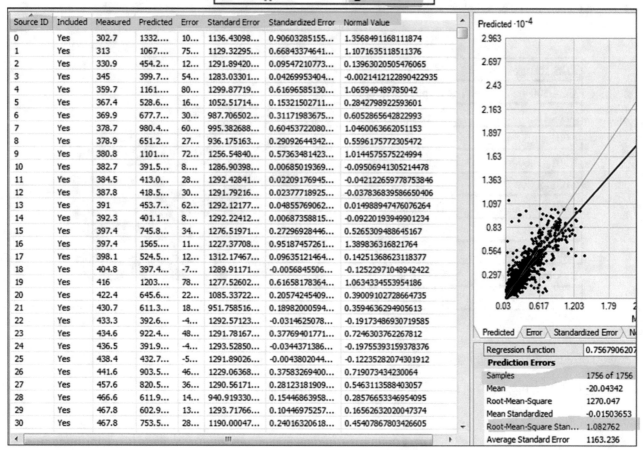

51. Click Finish then click OK

Result: Kriging Prediction Map is an output layer and has the same area extend as the Well layer. The TDS ranges between 302.7 and 29,632 ppm.

Convert the Kriging Predicted Map into ESRI Grid and Clip It Using the Mask Technique

52. Right click Kriging Prediction Map/Data/Export to Raster
53. Click the Environment Item
54. Open Workspace and make sure the Workspace and Raster Analysis are set as below:
    f. Current Workspace: \\Data\Q1
    g. Scratch Workspace: \\Result
55. Open Raster Analysis
    h. Cell Size: As Specified Below
    i. Type 50
    j. Mask Watershed
56. Click OK
57. Input geostatistical layer: Kriging Prediction Map
58. Output Raster: \\Result\Kriging
59. Output cell size: **50**
60. Click OK
61. In Model # 1 Select the Type "Spherical"
62. Anisotropy select True/Next
63. Accept the Default/Next

Classify the **Kriging** manually into 10 classes 500, 1000, 1500, 2000, 3000, 4000, 5000, 10000, 15000, and 30000 as in the IDW and match the color as below.

TDS Range	Color
332 - 500	Apatite Blue
501 - 1,000	Cretan Blue
1,001 - 1,500	Solar Yellow
1,501 - 2,000	Light Apple
2,001 -3,000	Leaf Green
3,001 - 4,000	Fir Green
4,001 - 5,000	Lilac
5,001 - 10,000	Ginger Pink
10001 - 15,000	Mars Red
15,001 - 30,000	Tuscan Red

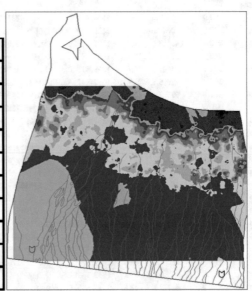

Question1: Comment on the salinity interpolation.

Question2: Run the GPI, IDW, and Kriging using the nitrate (NO_3) variable.

Watershed Delineation

Introduction

The watershed is an area of land that serves as a catchment for water. From the watershed, the surface water then enters a common outlet in the form of either a body of water, such as a lake, stream, or wetland; or it infiltrate into the groundwater. It is simply an area that drains surface water from high elevation to low elevation. The watershed is a hydrologic unit that is used to be modeled as it is considered fundamental to hydrologic designs and it is used to aid in the study of the movement, distribution, and quality and quantity of water in an area. The watershed analysis is a technique essential in the management, conservation, and planning of the Earth's natural resources.

Traditionally, the watershed is created mannually from topographic maps by locating the water divide. In ArcGIS the watershed can be delineated using Spatial Analyst extension and the hydrology tools for watershed delineation.

There are many steps involved in creating the watershed boundary. The delineation of the watershed requires work with the raster DEM of a study area. If the raster DEM is not available and you have a point elevation, you can use one of the interpolation techniques such as IDW, GPI, or kriging in ArcGIS to convert the point elevation into DEM. The watershed can be created using DEM as it is considered the main source point to create a watershed model.

After obtaining the DEM, make sure that the raster is depression free. These topographic depressions are also called sinks. Depression is normal in nature, and could be generated during the interpolation process of DEM creation. Depression in DEM is when a very low elevation relative to neighboring cells is found and they prevent downslope DEM flow-path routing. These low elevation cells can be removed by increasing its cell value to the lowest overflow point. The table below shows that the cell in row 3 column 2 is a depression.

450	446	441	451	454
447	441	440	446	451
440	339	431	440	445
438	435	422	431	439
431	429	414	422	424

Therefore, to use the raster DEM in watershed delineation, the depressions should be removed using the Fill tool.

Electronic Supplementary Material: The online version of this chapter (https://doi.org/10.1007/978-3-319-61158-7_14) contains supplementary material, which is available to authorized users.

Flow Direction

The next step is to create a raster grid containing the information about flow directions. The Flow Direction tool resides in the hydrology tool. Flow Direction tool is used to find drainage networks and drainage divides and it is determined by the elevation of surrounding cells in the DEM. The water can flow only into one cell and the GIS model assumes no sinks. If this does not happen in the output grid, this means that the raster DEM has sinks. Flow direction is critical in the hydrologic modeling process because it determines the direction of flow for each cell in the land topography. The raster grid that created by the Flow Direction tool is based on the D8 flow algorithm. The D8 algorithm is the method for performing the flow path analysis for the application of watershed delineation. The method will assign a cell's flow direction to one of its eight surrounding cell that has the steepest gradient. D8 has disadvantages and can be replaced by D∞. Flow direction function is that for each 3 × 3 cell neighborhood, the grid processor stops at the center cell and determines which neighboring cell is lowest. Depending on the direction of flow, the output grid will have a cell value at the center cell. The values for each direction from the center are 1, 2, 4, 8, 16, 32, 64, and 128. For example, if the direction of steepest drop was to the left of the current processing cell, its flow direction would be coded as 16. The figure below shows the output grid cell values with the center cell, as determined by this matrix:

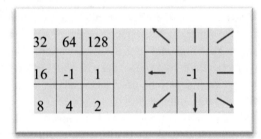

If a cell flows northwest, then in the output grid, the cell in its location will have a value of 32 and if a cell flows southward, then the value will be 4 (figure below)

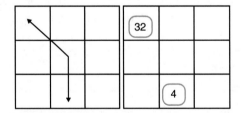

Flow Accumulation

The next step in creating the watershed is to run the Flow Accumulation function, which is an important step to create the drainage network and measure the area of a watershed that contributes runoff to a given cell. Therefore, it is a necessity to determine the ultimate flow path based on the direction of flow of every cell on the topography grid. Flow accumulation selects cells with the greatest accumulated flow, which will assist in creating a network of high-flow cells. These high-flow cells should be situated on stream channels and at valley bottoms. Cells that have high accumulation values, higher than "**1**" correspond to stream flow and cells having an accumulation value of "0" correspond to ridgelines. Once flow accumulation is calculated, it is customary to identify those cells with high flow. Higher-flow cells will have a larger value and the user can select any threshold number (i.e. >500), which should be close to network obtained from the traditional method.

Stream Link

After the stream network is established from flow accumulation, each stream section of the stream raster is assigned a unique value e.g. 1, 2, 3, etc. The intersection of the streams is like nodes and the stream section is the arc.

To delineate a watershed, you need to select an outlet cell (Pour Point), which is the lowest point in the watershed, where all flow is directed. The Pour Point also can be any feature such as a gauge station, dam, sampling location, confluence of a tributary with a main stream or any point of interest in the study area. The Pour Point could be a raster or vector. To obtain the watershed, the Pour Point should coincide with the flow path of high flow accumulation values in the flow accumulation raster. Finally, you can use the Watershed tool to extract the whole watershed polygons for the Pour Points, or a single watershed for a specific stream or tributary. At the end, you can convert your raster watershed to vector so that you can integrate it and align it nicely with the rest of your digital data.

Scenario 1: You are a geologist working for WAJ and one of your duties is to delineate the catchment area of the Dhuleil-Halabat region. This catchment is going to be used as an input function to estimate the recharge amount to the groundwater resources in the area.

Your duty is to do the following:

1. Delineate the watershed based on the whole region.
2. Delineate the watershed based on a single point (Pour Point)
3. Convert the watershed raster into vector
4. Calculate the area of the watershed and amount of groundwater recharge

Required Files: **DEM, Dhuleil.shp,** and **Pourpoint.shp**

1. Launch ArcMap and rename the Layers data frame "Dhuleil"
2. Add Data, browse to \\Ch14\Data folder and integrate **Dhuleil.shp**, **Pourpoint.shp**, and **DEM**
3. Make the symbol of **Dhuleil.shp** hollow and the outline color red
4. Symbolize **Pourpoint.shp** (circle 2, size 12, and blue color)

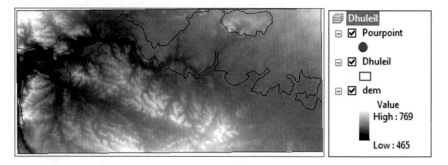

Delineate the Watershed

Delineating the watershed requires different steps:

Step 1: Run the Flow Direction Tool

This step is very important to run in order to create a raster grid containing the information about flow directions and identify any depressions in the raster DEM. The DEM raster file will be used for performing the task

5. Launch ArcToolbox/Set Environment
6. Current Workspace: \\Ch14\Data
7. Scratch Workspace: \\Result
8. OK

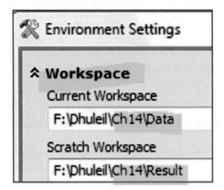

9. Spatial Analyst Tools/Hydrology/Flow Direction
    a. Input surface raster: **dem**
    b. Output flow direction raster: **Flowdirec.tif**
10. Ok

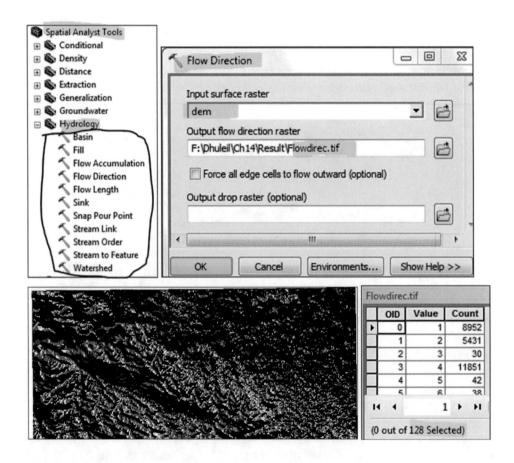

Result: The output raster is an integer and contains 128 records. The grid has no information about the flow directions, but has recognized that the **DEM** contains sinks.

Step 2: Identify the Locations of the SINK (Sink Tool)

1. Spatial Analyst Tools/Hydrology/Sink
2. Input flow direction raster: **Flowdirec.tif**

3. Output raster: **Sink.tif**
4. Ok

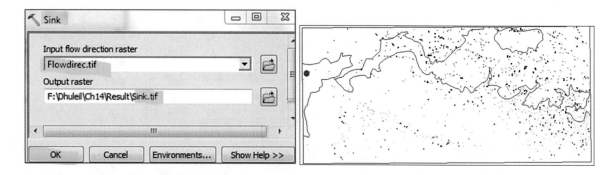

Comments: The output grid shows that the DEM of Dhuleil has 927 **sinks**. A sink is a depression or raster cell surrounded by higher elevation values. Some sinks are actually in nature and some are deficiencies in the DEM. After knowing the number of sinks, we have to fill them using the "Fill tool" on the original DEM raster.

Step 3: Run the Fill Tool

5. Spatial Analyst Tools/Hydrology/Fill
   a. Input surface raster: **dem**
   b. Output surface raster: \\Result\ DEM_Fill.tif
6. OK

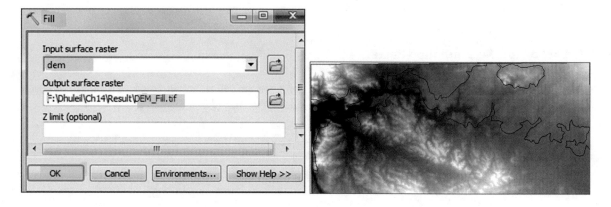

Result: Now the **DEM_Fill.tif** is a DEM free of any SINKS.

Step 4: Run the Flow Direction tool

7. Spatial Analyst Tools/Hydrology/Flow Direction
   a. Input surface raster: **DEM_Fill.tif**
   b. Output flow direction raster: \\Result\Flowdirection.tif
8. Ok

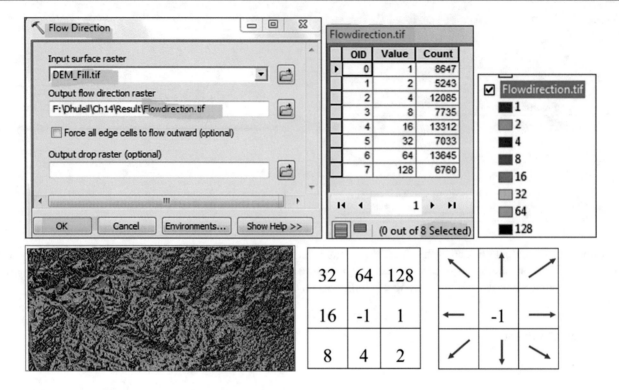

Result: The output Flowdirection.tif raster is an integer raster with the attribute table of eight records.

Comments: The values 64 and 16 have the highest frequency, which means the direction of the surface flow is north and west.

Step 5: Create a Flow Accumulation Raster

This step is important as it will tabulate for each cell the number of upstream cells that will flow into it and the tabulation will be based on the flow direction raster. This step will derive two values; "**0**" and "**1**".

9. Spatial Analyst Tools/Hydrology/Flow Accumulation
   a. Input flow direction raster: **flowdirection.tif**
   b. Output accumulation raster: \\Result\flowAccum.tif
   c. Output Data Type: INTEGER
10. Ok

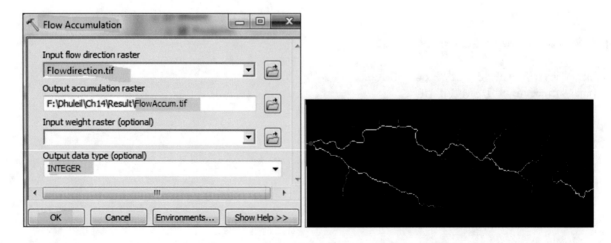

Question: What is the range of cell value in **flowAccum.tif**?

Step 6: Create Source Raster to Delineate Watershed

The flow accumulation raster will help in deriving the stream network. The stream derivation is based on threshold cell values, which could be 100, 500, or more cells. The 500 cells means that each cell has a minimum 500 cells contributing to them. The difference between the 500 cells and the 100 cells is the 500 cells will generate less dense stream networks, while the 100 cells will generate denser streams.

Source Raster

The source raster required two steps:

First Step: This step will allow you to select the cells in the flow accumulation raster that have more than 500 cells flowing into them.

11. Spatial Analyst Tools/Conditional/Con
    a. Input Conditional raster: FlowAcum.tif
    b. Expression (SQL) "Value" > 500
    c. Input true raster or constant value: 1
    d. Output raster: \\Result\Network.tif
12. OK

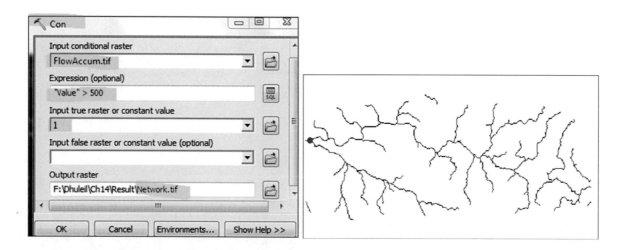

Result: The Network raster has only 1 record.

Second step: Assign a unique value to each section of the **Network** raster and its associated flow direction. In other words, make each segment where it intersects with another segment as a unique record.

13. Spatial Analyst Tools/Hydrology/Stream Link
    a. Input stream raster: **Network.tif**
    b. Input flow direction raster: **flowdirection.tif**
    c. Output raster; \\Result\StreamLink.tif
14. Ok

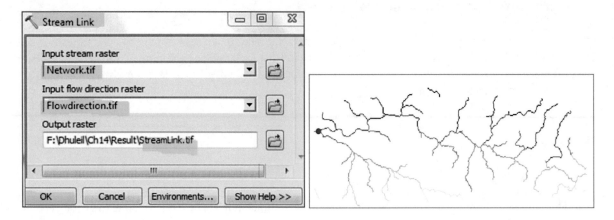

Result: The StreamLink raster has 82 segments and each stream segment is now an independent record and has a "Value" in the attribute table.

Step 7: Delineate Watershed

15. ArcToolbox/Spatial Analyst Tools/Hydrology/Watershed
    a. Input flow direction raster: **flowdirection.tif**
    b. Input raster or feature pour point data: **StreamLink.tif**
    c. Output raster; \\Result\Watershed
16. Ok
17. D-click **Watershed** /Symbology/ Color ramp /Elevation 1/OK

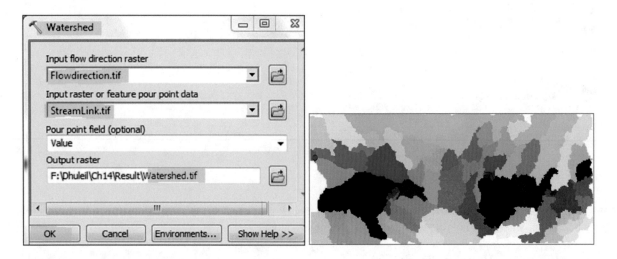

Result: Watershed is created.

Question: How many sub-basins are in the watershed?

Point-Based Watershed

This approach allows the user to derive a watershed either for each stream or based on a point of interest such as any point of interest along the flow system.

Scenario 2: Your superior asked you to create only one watershed based on a point (Pourpoint.shp) and then calculate the total area of the watershed.

Run the watershed using **Pourpoint**

18. Insert Data Frame and rename it to **PourPoint**
19. Drag the **Flowdirection.tif**, and **Pourpoint.shp** from Dhuleil data frame
20. Spatial Analyst Tools/Hydrology/Watershed
21. Input flow direction raster: **Flowdirection.tif**
22. Input raster or feature pour point data: **Pourpoint**
23. Output raster: **Pourshed.tif**
24. OK
25. Change the color ramp of the **Pourshed.tif** into Blue Bright

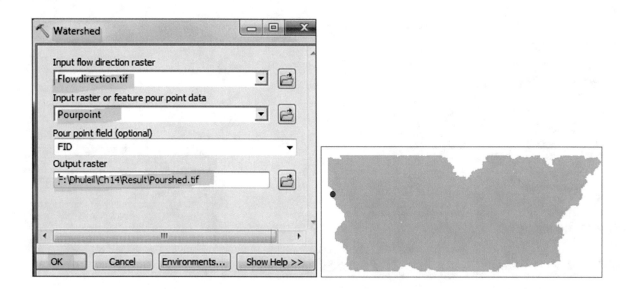

Convert Pourshed Raster to Vector

26. ArcToolbox/Conversion Tools/From Raster/Raster to Polygon
27. D-click Raster to Polygon
28. Input raster: Pourshed.tif
29. Output polygon features: \\Result\Dhuleil_Watershed.shp
30. Ok

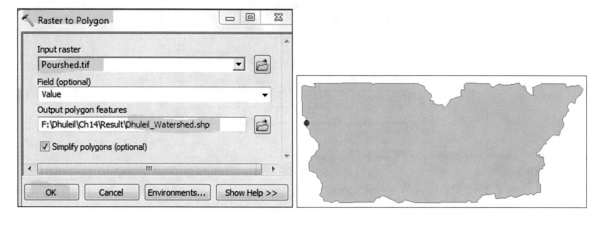

Calculate the Recharge amount of the Dhuleil_Watershed

In order to calculate the Dhuleil_Watershed area, you have to project it into plane projection (Palestinian Grid)

31. ArcToolbox/Data Management Tool/ Projections and Transformations/Feature
32. D-click Project and fill as below
33. Input Dataset: Dhuleil_Watershed
34. Input Coordinate System: GCS_WGS_1984 (Default)
35. Output Dataset: \\Result\Watershed_TM.shp
36. Output Coordinate System: Browse\Projected Coordinate Systems\National Grid \Asia\Palestine 1923 Palestine Grid
37. OK
38. Geographic Transformation: Palestine_1923_T0_WGS_1984_1
39. OK

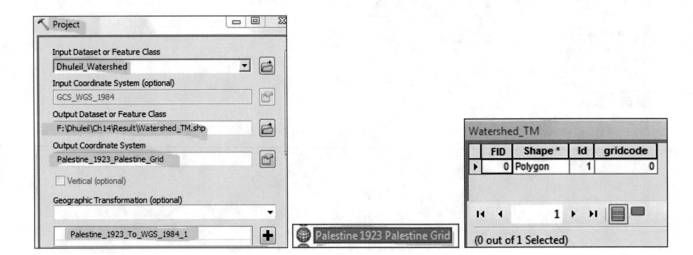

40. Open attribute table of Watershed_TM
41. Table Option/Add Field
42. Name: Area
43. Type: Double (Precision 12, Scale 2)
44. OK

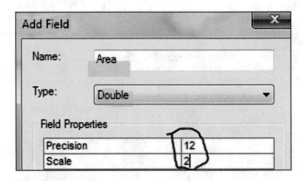

45. R-click Area/Calculate Geometry/click Yes
46. Property: Area
47. Coordinate System: Palestine 1923 Palestine Grid
48. Units: Square Meters

49. OK

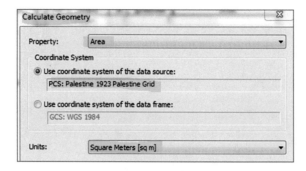

50. What is the total area?

Task: Calculate the recharge, if the amount of precipitation is 120 mm and the infiltration rate is 5%.
Hint:

1. Recharge = Area (m^2) × Infiltration Rate (120 × 0.05 mm)
2. Convert the mm into meter (1 mm = 0.001 m)

Introduction

Geostatistics is a very useful approach that allows users to obtain meaningful information related to data in term of its distribution and patterns in GIS. This chapter includes some applications of Spatial Statistics from the GIS environment based on groundwater data. The intention is to focus on the application of GIS rather than emphasizing on complex mathematical and statistical theories. Nevertheless, some of the tools such as Measuring Geographic Distribution, Analysis Patterns, and Mapping Clusters of the Spatial Statistical analysis will be explained and applied using groundwater data.

Measuring Geographic Distribution Toolset

Using geographic distribution tools in ArcGIS aiming to perform statistical approaches to assist researchers in measuring the distribution of features. The tools allow users, for example, to calculate a value that represents a characteristic of the distribution. Such as the center of groundwater wells tapping an aquifer. By doing this, you can see how the wells are dispersed throughout the basin. There are three types of centers that can be calculated: Mean Center, Median Center, and Central feature.

1. **Mean Center**: is the average of the X-coordinate and Y-coordinates values of all features. The resulting X, Y coordinate pair is the mean center. For example in the Jarash area, there are several wells (Figure below) that spread through the area and in order to find the mean center, we calculate the averages of both X and Y coordinates (table below).

Electronic Supplementary Material: The online version of this chapter (https://doi.org/10.1007/978-3-319-61158-7_15) contains supplementary material, which is available to authorized users.

ID	Depth	TDS	X-Coordinate	Y-Coordinate
1	376	1307	228970.98	1189834.73
2	200	4160	232948.09	1185966.58
3	123	3965	236816.24	1191033.32
4	90	1950	237796.90	1180899.85
5	150	2015	231640.55	1176868.26
6	60	1872	219545.77	1179210.94
Average			231286.42	1183968.95

2. **Central feature**: is the feature that associated with the smallest accumulated distance to all other features in a study area. For example there are 6 wells in the Jarash area (Figure below) and to calculate the central feature, the 6 wells (table below) will be organized into a table. The 6 wells are represented as records and columns and then the distance between the wells will be recorded. The sum of total distance of each well from the rest of the wells is then recorded, and the central feature will be the well that has the lowest total distance from all other wells. In the table below, you can see that well No 2 is selected as the central feature.

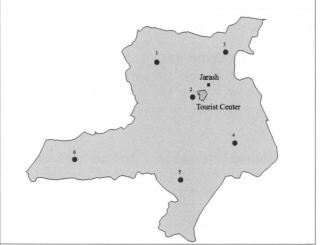

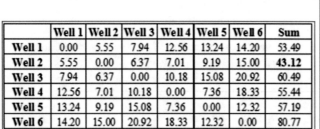

	Well 1	Well 2	Well 3	Well 4	Well 5	Well 6	Sum
Well 1	0.00	5.55	7.94	12.56	13.24	14.20	53.49
Well 2	5.55	0.00	6.37	7.01	9.19	15.00	**43.12**
Well 3	7.94	6.37	0.00	10.18	15.08	20.92	60.49
Well 4	12.56	7.01	10.18	0.00	7.36	18.33	55.44
Well 5	13.24	9.19	15.08	7.36	0.00	12.32	57.19
Well 6	14.20	15.00	20.92	18.33	12.32	0.00	80.77

3. **Median Center** Median Center is a slightly different way to calculate the middle (actually quite complicated to compute) and is a point in a pattern which minimizes the distance between itself and all other points. Median Center identifies the location that minimizes overall Euclidean distance to the features in a dataset.

Calculate Mean Center with and Without Weight

This section will explore the Mean Center with or without weight.

Mean Center

In order to find the center of a random distributed features over an area, you need to use the Mean Center tool resides in the spatial statistics tools. Calculating the center has many applications in applied sciences, especially in geoscience. The center is a feature in the middle of a given set of data and can service all other feature with a shortest time. For example, a set of groundwater wells is located in a particular study area, and finding the center of the wells will help building a water tower that will collect water from the surrounding wells faster and with less expense.

Scenario 1: In Jordan, in summer time, the demand for potable water increases. The Water Authority in the Jarash governorate has decided to build an extra water tower in the area to be used as a distribution center during the summer. The water tower should be supplied with water from a high quality groundwater source and it should be located in the center of the wells that belong to the major cities in the governorate.

GIS Approach

1. Launch ArcMap and call the Layers Data Frame "**Mean Center**"
2. Integrate from \\Ch15\Data\Q1 folder **Governorate.shp**, **Town.shp**, and **Well.shp**
3. Click the symbol of Town layer in the TOC, select Square 2, color pink, size 9, OK
4. D-click Well layer in TOC/Symbology tab/choose categories and unique values/Value Field, enter "city"/click on "Add All Values"/uncheck <all other value>
5. Click on "Symbol" and choose properties for all symbols/choose circle 2 and change the size to 10/OK/choose Basic Random from the Color Ramp/OK
6. Click the symbol of Governorate/click Hollow and then OK

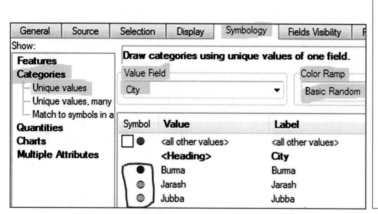
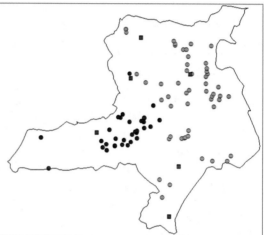

7. Launch ArcToolbox, click an empty place at the bottom of ArcToolbox
8. Click Environment/Workspace
9. Current Workspace \\Ch15\Data\Q1
10. Scratch Workspace \\Result
11. OK

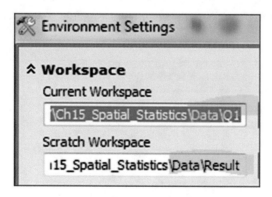

12. ArcToolbox/Spatial Statistics Tools/Measuring Geographic Distributions
13. D-click Mean Center
14. Input Feature Class: Well
15. Output Feature Class: \\Result\Well_Mean_Center.shp

16. Case Field: City
17. OK
18. D_click on **Well_Mean_Center** layer/symbology tab/Categories/unique values/Choose "City" as the value field/click on add all values. Click on "symbol" and "properties for all symbols"/Change the symbols to hexagon 2, make the size as 20, then OK. Choose different colors to match the color of the wells in each town.

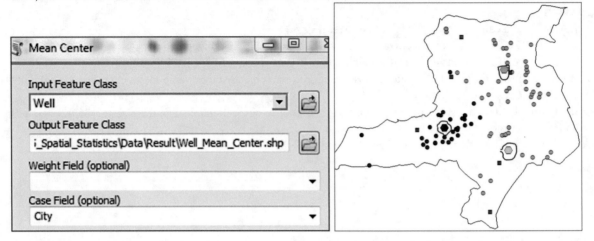

Mean Center with Weight

The Next Step is to run the **Mean Center** again on the well layer using the **Weight** field. The **Weight** field has values from 1 to 2. Value 2 indicates the most important criteria, and signifies wells that have TDS and NO_3^- less than 1000 and 45 mg/L respectively and depth of wells less than 300 m.

Run the **Mean Center** function again using the **Well** layer and the **Weight** field, call the output **Well_ Weight_Mean_ Center.shp**

1. D-click Mean Center
2. Input Feature Class Well
3. Output Feature Class \\Result\Well_Weight_Mean_Center.shp
4. Weighted Field: Weight
5. Case Field: City
6. OK
7. D-click on "Well_Weight_Mean_Center layer/Symbology/Categories, unique values" and again choose the city as the value field.
8. Click add all values and click on symbol and choose properties for all symbols.
9. Change the symbol to hexagon 2 size 18/choose the color ramp as before so the well weight mean center has the same color scheme as the cities.

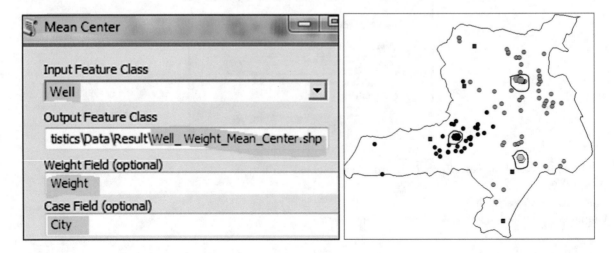

Questions:

1. Are the **Well_ Weight_Mean_Center.shp** aligns with the **Well_Mean_Center**?
2. If they don't align how far are they from each other?
     Hint: Use the "measure" tool to measure the distance between the mean center and weighted mean center for each wells in the three cities.
3. Run the Central Feature tool and compare the result with the result of the mean center and comment on the result

Standard Distance and Mean Center

The standard distance calculate the mean center of the displayed features and then draws a buffer around the mean center with a radius equal to the standard distance value. There are three values of the standard deviation. The first standard deviation will cover at least 68% of the sample features, the second standard deviation cover at least 95% of the sample features, while the third standard deviation covers almost 99% of all the samples.

Scenario 2: You are a hydrogeologist and your goal is to replenish the groundwater and improve its water quality using an artificial recharge method. To perform this task, you decided to at least select one well that is located within 1 standard deviation from the mean center and in close proximity to the Khaldiyah dam. To execute the assignment, you decided to calculate the standard circle using a "**weight**" criteria. The weight is based on total depth of the wells and more emphasis is placed on the wells that are shallower than 100 m. A field called weight is added to the attribute table of the wells.

GIS Approach
1   Insert Data Frame and call it Standard Distance
2. Integrate **Dam.shp**, **Geology.shp**, **Stream.shp**, and **Well.shp** from \\Ch15\Data\Q2 folder
3. D-click the Geology layer in the TOC/Symbology/Categories/Unique values = Code/Add All Values/Uncheck <all other values/OK
4. Click the symbol of the Stream in TOC/select the River symbol in the Symbol Selector dialog box/then click OK
5. Click the symbol of the Dam in TOC/select the Dam Lock symbol in the Symbol Selector dialog box/then click OK (or search for "Dam Lock" in the Symbol Selector)
6. Click on "Symbol" of the Well in TOC/choose circle 2 and change the size to 10/choose blue color/then click OK

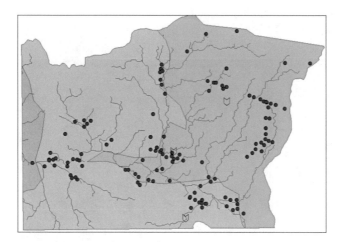

7. ArcToolbox/Spatial Statistics Tools/Measuring Geographic Distributions
8. D-click Standard Distance
9. Input Feature Class: Well
10. Output Standard Distance Feature Class: \\Result\Well_StndDist.shp
11. Circle Size: 1_Standard_Deviation
12. Weight Field: Weight
13. OK

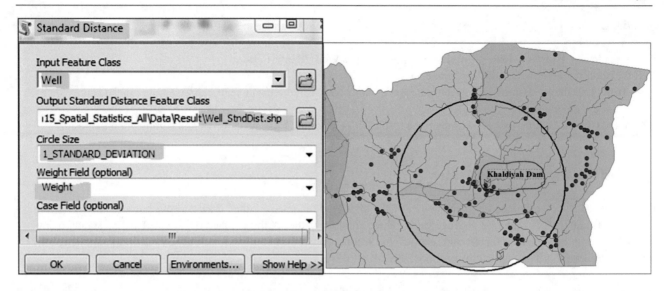

Result: The **Standard distance** will calculate the mean center based on the weight, and then a buffer is created around the mean center using 68% of the wells in the study area.

Note: In order to pursue with the analysis you have to answer the following questions:

1. How far the Khaldiyah dam from the center of the buffer
2. What is the closest well to the dam that can be used in the artificial recharge

Distance of the Khaldiyah Dam from the Center of the Buffer

To calculate the distance of the Khaldiyah Dam from the center of the buffer, you have to run the Mean Center tool as shown above and then use the measure tool to find the distance between the center of the buffer and the dam.

14. D-click Mean Center
15. Input Feature Class: Well
16. Weight Field: Weight
17. Output Feature Class: \\Result\Mean_Center_Dam.shp
18. OK

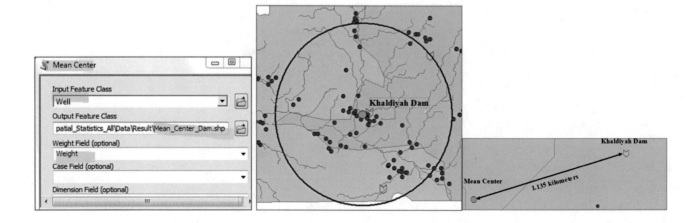

In order to find the well that is close to **Khaldyia dam**, you have to zoom in around the **Khaldyia dam** and use the measure tool to see, which well is the closest.

Result: It is clear that the Well No 109 has the shortest distance (419.88 m) to the **Khaldyia dam**.

Analyzing Pattern Toolset

Identifying Pattern Based on Location

Some statistical analyses aims to identify patterns, trends, and spatial relationship among features in any environment. Whether a certain set of data is more likely to show certain characteristics; Some Spatial Analyst tools are able to recognize the distribution patterns of geographic layers in a specific study area. In the geography Discipline there is a well documented practice that demonstrates how features are located near each other are more similar than features situated farther away from one another (Tobler's First Law of Geography). This idea is common sense, nevertheless, there is always exception to the rules. For example the weather in the Jordan Rift Valley, which is about 400 m below sea level is not similar to weather in the Ajloun High Lands, which is more than 1000 m above sea level and is only 20 km far away from each other. At the same time, the climate of the city of Aqaba (Jordan) is similar to the city of Jeddah (Saudi Arabia), even if the two cities are 970 km away from each other.

Average Nearest Neighbor tool

The "*Average Nearest Neighbor*" tool will detect if features are clustered or dispersed; this tool is used and tested with some degree of confidence level. The statistical approach behind this method is that the tool will measure the distance from each feature in the dataset to its single nearest feature neighbor and then calculating the average distance of all measurements. The tool then creates a hypothetical dataset with same number of features, but placed **randomly** within the **study area**. The tool then will be run again and measure the nearest distance to its nearest neighbor feature and calculate the average. The average distance of the random hypothetical data will be assessed with the real data. Two parameters will be generated: I and Z-score

A: a nearest neighbor index (I) is generated as follow

$$I = \frac{D_r}{D_h}$$

D_r is the calculated average distance of the real data
D_h is the average distance from the hypothetical data

> If I < 1 the data show clustering
> If I > 1 the data show dispersion
> If I = 1 the data is randomly distributed

A pattern that falls at a point between dispersed and clustered is said to be random

B: A z-score will be calculated and is vital to making a decision to accept or reject the null hypothesis. The z-score is associated with the confidence level and is up to the researchers to adopt which confidence level they are willing to test with their hypothesis. Each confidence level is associated with z-score which is simply a standard deviation. For example, a 90% confidence level has a z-score between −1.65 and +1.65 and the 95% confidence level has a z-score between −1.96 and +1.96 (table below)

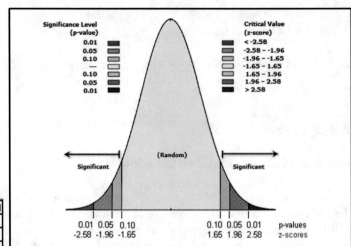

Z-score (Standard Deviations)	p-value (Probability)	Confidence level
-1.65 or +1.65	0.10	0.9
-1.96 or +1.96	0.05	0.95
-2.58 or +2.58	0.01	0.99

Null Hypothesis

In any statistical testing, you have to propose a **null hypothesis** and the null hypothesis states that features in the study area lacking any pattern. This means that the features are not clustering or dispersed, but randomly distributed.

Let us assume you are willing to test your hypothesis with 95% confidence level and you are assuming (Null Hypothesis) that the features are **randomly distributed**. After running the test the Z-score value that generated is between **−1.96 and +1.96** and your p-value is larger than **0.05**. Based on the result you have to **accept** the null hypothesis, which means that your features are randomly distributed. But if the Z-score fell outside that range for example −2.0 or +2.0 standard deviations, you have to **reject** the null hypothesis and your observed features are clustering or dispersed.

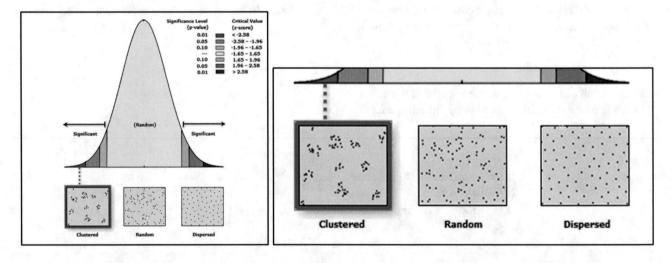

Scenario 3: You are a hydrogeologist and you have observed a heavy groundwater abstraction from the wells that are used for irrigation in Wala catchment area. This practice has dramatically lowered the water table in these wells, which affected the water balance in the whole basin. You have decided to examine if the wells' locations are one of the reasons that generate a dramatic dropdown. The nearness of the wells from each other could affect the zone of influence created by the wells' pumping. Your question is, are the wells used for agriculture randomly distributed or do they have a certain pattern (clustered or dispersed). To find out the answer you have to do the following:

GIS Approach

Propose a *Null Hypothesis* stating that wells are randomly distributed in the basin

1. Insert Data Frame and call it "**Nearest Neighbor**"
2. Integrate the **Well.shp** and **WalaWatershed.shp** into from \\Ch15\Data\Q3 folder
3. R-click Well layer/Open Attribute Table (there are 333 wells)
4. R-click Well layer in the TOC/Properties tab/click Definition Query/click Query Builder
5. Write the SQL statement: "Type" = 'Irrigation'
6. OK/OK

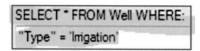

Result: The well layer that used for irrigation is the only one is displayed and the rest of the wells are hidden now. You can verify that by opening the attribute table.

7. R-click Well layer/Open Attribute Table (there are 271 wells)
8. Open attribute table of WalaWatershed, the Shape_Area is **1,803,591,128.58 (unit is m²)**

	FID	Shape *	OBJECTID	CATCH_DESC	Shape_Leng	Shape_Area
▶	0	Polygon	37	Wala Watershed	208,730.76	1,803,591,128.58

WalaWatershed

9. ArcToolbox/Spatial Statistics Tools/Analyzing Patterns
10. D-click "Average Nearest Neighbor"
11. Input Feature Class: Well
12. Distance Method: EUCLIDEAN_DISTANCE
13. AREA 1803591128.58
14. Check "Generate Report"
15. OK
16. Click Geoprocessing menu/and point to Results
17. Open the Current Session
18. Open Average Nearest Neighbor

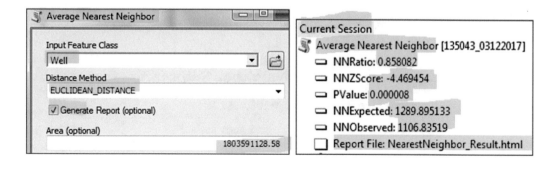

19. Write down the following values:

> Observed Mean Distance (NNObserved) = 1106.84
> Expected Mean Distance (NNExpected) = 1289.89
> Nearest Neighbor Ratio (NNRatio) = 0.858
> Z-Score: (NNZScore) = −4.47

20. D-click Report File: Nearest Neighbor_Result.html
    Result: The following graph displays

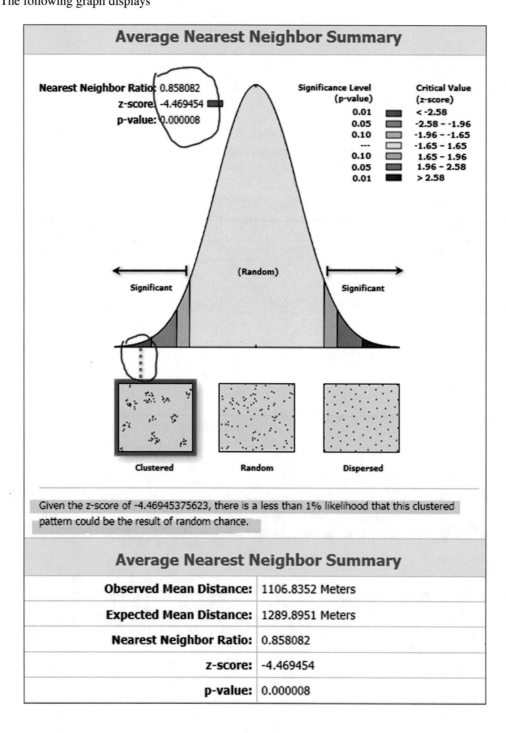

Average Nearest Neighbor Summary	
Observed Mean Distance:	1106.8352 Meters
Expected Mean Distance:	1289.8951 Meters
Nearest Neighbor Ratio:	0.858082
z-score:	-4.469454
p-value:	0.000008

Interpretation: Given the z-score of −4.47, there is less than 1% possibility that the irrigation wells in the Wala catchment area could be the result of random chance. Because the "I" ratio is also 0.858, which is less than 1, we reject the null hypothesis and we consider the distribution of the irrigation wells clustered.

Identify Pattern Based on Values (Getis-Ord General G)

The location of features is not only determined by the clustering but also the values associated with the feature within a key distance of each other. The General G-statistics tool will be used to identify high or low values over the entire study area. The distance, which is based on either Euclidean or Manhattan will be calculated and it will reveal whether if it is significant or not. The tool allows user also to specify how spatial relationship among features are defined. For example, the "Fixed Distance Band" Each feature is analyzed within the context of neighboring features. Neighboring features inside the specified critical distance (Distance Band or Threshold Distance) receive a weight of one and exert influence on computations for the target feature. Neighboring features outside the critical distance receive a weight of zero and have no influence on a target feature's computations. The distance is an important part of the General G-statistics as it will show over which the tool will be ascertained to be significant. The ideal distance will be determined using the "Calculate Distance Band from Neighbor Count" tool.

The "Calculate Distance Band from Neighbor Count" tool returns the minimum, the maximum, and the average distance to the specified Nth nearest neighbor (N is an input parameter) for a set of features, for example 5 wells.

The General G tool calculates the value of the General G index, Z score and p-value for a given input feature class. The Z score and p-value are measures of statistical significance which tell you whether or not to reject the null hypothesis. For this tool, the null hypothesis states that the values associated with the features are randomly distributed. The Z score value means the following:

a) A Z score near zero indicates no apparent clustering within the study area.
b) A positive Z score indicates clustering of high values.
c) A negative Z scores indicates clustering of low values.

Scenario 4: In the previous scenario we found out that the irrigation wells are clustering, this time you have to see if the "**Weight**" field has an influence on the clustering of the same wells in the watershed and at what distance the clustering taking place. The "**Weight**" field has values from 1 to 5, with 5 representing the wells that have the highest yields. In this exercise, you have to run the "Calculate Distance Band from Neighbor Count" tool to find the ideal distance to run the General G-statistics. This will show over the tool which will be ascertained to be significant. After identifying the average distance which will return the minimum, the maximum, and the average distance to the 5th nearest neighbor (N = 5 wells), you should use value higher and lower than the average return value and run all of them to decide which distance is ideal to use the General G-statistics.

1. Insert Data Frame and call it "**General G-Statistics**"
2. Integrate the **Well.shp** and **WalaWatershed.shp** from \\Ch15\Data\Q4 folder
3. R-click Well layer/Open Attribute Table (there are 333 wells)
4. R-click Well layer in the TOC/Properties tab/click Definition Query /click Query Builder
5. Write the SQL statement: "Type" = 'Irrigation'
6. OK/OK

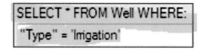

SELECT * FROM Well WHERE:
"Type" = 'Irrigation'

Result: The well layer that used for irrigation is the only one is displayed and the rest of the wells are hidden. You can verify that by opening the attribute table.

Null Hypothesis: Irrigation wells with high-ranking values represented by the "**Weight**" field are randomly distributed in the study area.

Find the Ideal Distance

Before running the General G-statistics, you have to run the "Calculate Distance Band from Neighbor Count" tool in order to find the best distance to use with the General G-statistics.

1. ArcToolbox/Spatial Statistics Tools/Utilities
2. D-click Calculate Distance Band from Neighbor Count
3. Input Features: Well
4. Neighbors: 5
5. Distance Method: EUCLIDEAN DISTANCE
6. OK

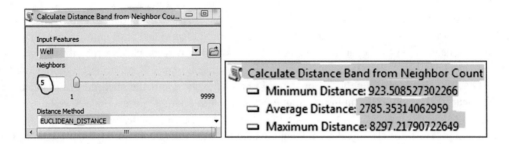

Result: If you open Geoprocessing menu/Result, you will notice the following:

1. Minimum Distance: 923.50
2. Average Distance: 2785.35
3. Maximum Distance: 8297.21

The average distance is 2785 m with five neighbors, therefore a lower and higher number should be used to determine which value to use to run the General G statistical tool. The distance to try should be from 2000 to 3400 m at 400 m intervals. Run all these values using the "High/Low Clustering (Getis-Ord General G)" tool and fill the table at the end.

7. ArcToolbox/Spatial Statistics Tools/Analyzing Patterns
8. D-click "High/Low Clustering (Getis-Ord General G)"
9. Input Feature Class: Well
10. Input Field: Weight
11. Check the Generate Report
12. Conceptualization of Spatial Relationships Fixed_Distance_Band
13. Distance Method: Euclidian_Distance
14. Standardization: None
15. Distance band: 2000
16. OK

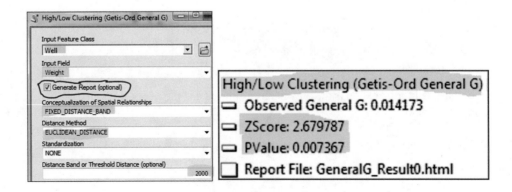

17. Click Geoprocessing menu/and point to Results
18. Open the Current Session
19. You see the result of the analysis
20. D-click on the Report File: GeneralG_Result

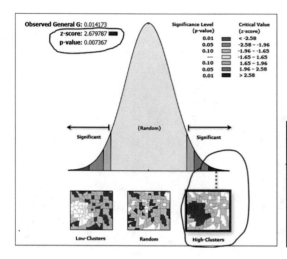

Distance (m)	Observed General G	Z-Score
2,000	2.679	0.007
2400	1.25	0.018
2800	1.32	0.024
3000	1.34	0.027

21. Repeat the previous step and replace the distance with 2400, 2800, 3000, and 3400

Interpretation: The best distance for clustering is the one that has the highest z-score.

Spatial Autocorrelation (Moran's I)

Moran's I index measure the spatial correlation using the feature location and an attribute value together to determine statistically if the data is clustered, dispersed or random. Using the spatial correlation helps define how the variables are arranged in a study area. The tool calculates the Moran's I Index value, Z score, and p-value. If the Moran's Index value is near +1.0 this indicates clustering while an index value near −1.0 indicates dispersion. The method has no output layer, but a report is established and demonstrate whether the well distribution in the watershed is clustered, dispersed, or random.

The tool will be run using the conceptualization of spatial relationship of Zone of indifference, Euclidian distance, and distance band of 500, 1000, 1500, 2000, 2500, and 4000. The concept of Zone of indifference that wells within the specified critical distance (Distance Band or Threshold Distance) of a target well receive a weight of one and influence computations for that well.

Scenario 5: Your supervisor asked you to look at the density of groundwater wells per block and would like to hear your judgment about at what distance these well densities cluster. This information is critical for management purposes as it helps to adjust the rate of pumping of the wells that are located close to each other in clustering pattern. Your duty is to do the following:

Prepare the Data for Analysis
1. Insert Data Frame and call it "**Moran's I Index**"
2. Integrate the **Grid_1000.shp**, **Well.shp**, and **WalaWatershed.shp** from \\Ch15\Data\Q5 folder

Spatial Join Between Grid_1000 and Well Layers
3. R-click Grid_1000/Join and Relate/click Join
4. Join Data from another layer based on spatial location
5. Choose the layer to join to this layer Well

6. Check "Each polygon will be given a summary of the numeric attributes of the points that fall inside it".
7. Save it in \\Result as **Well_Grid.shp**
8. OK

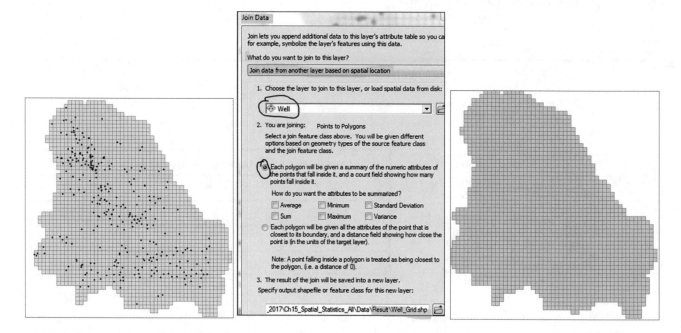

Result: The Well_Grid is created and it shows the density of the groundwater wells in the Wala watershed. If you open the attribute table, you will find a field called "**Count_**". Some of the values are zero, which shows the cells that has no wells. These cells should be removed before running the statistics.

9. R-click Well_Grid/Properties/Definition Query/click Query Builder and type the following statement "Count_" > 0
10. OK/OK

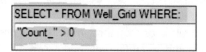

Result: The Well_Grid displays showing only the cells in the grid that have wells inside them. Some cells contain 1 well and other more than one well. The maximum wells found in one cell is 7, and they are in the northern part of the watershed.

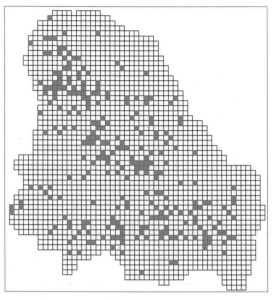

Well_Grid							
	FID	Shape	Well_FID	PageNumber	Area	Length	Count_
▶	33	Polygon	33	34	1000000	4000	1
	92	Polygon	92	93	1000000	4000	1
	93	Polygon	93	94	1000000	4000	2

Run Spatial Autocorrelation (Moran's I)

11. ArcToolbox/Spatial Statistics Tools/Analyzing Patterns
12. D-click "Spatial Autocorrelation (Morans I)"
13. Input Feature Class: Well_Grid
14. Input Field: Count_
15. Check the Generate Report
16. Conceptualization of spatial relationship: ZONE_OF_INDIFFERENCE
17. Distance Method: EUCLIDEAN DISTANCE
18. Distance Band: 500
19. OK
20. Click Geoprocessing menu/and point to Results
21. Open the Current Session
22. Open Spatial Autocorrelation (Moran I)

Result: The Spatial Autocorrelation tool returns the Moran's I Index, ZScore, and PValue, and the pattern. If you checked the Generate Report in the Spatial Autocorrelation (Moran I) tool, an HTML file with a graphical summary of results will be created.

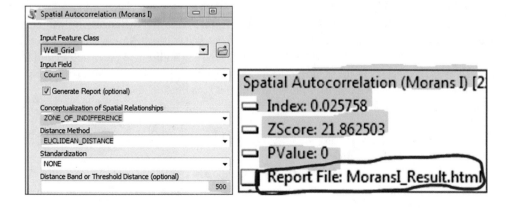

23. D-click on the HTML file, it will open the HTML file in the default Internet browser
24. Repeat the steps above and use 1000, 2000, 4000, 6000, and 8000 and fill the table below

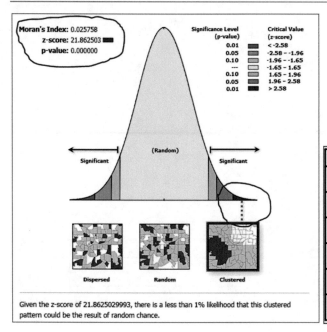

Distance	Z-Score	Moran's Index (I)	Pattern
500	8	0.066	Clustered
1000	3.43	0.289	Clustered
2000	5.03	0.241	Clustered
4000	5.87	0.147	Clustered
6000	7.94	0.134	Clustered
8000	7.62	0.098	Clustered

Interpretation: The most significant clustering occur at a distance, where the Z-score is the highest and the Moran's Index (I) is the lowest.

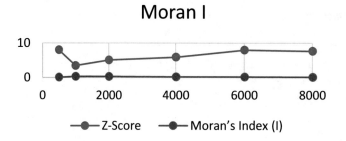

Mapping Clusters

Cluster and Outlier Analysis (Anselin Local Moran I)

The cluster analysis will examine a dataset of features (such as wells) with a value associated with the features (such as depth or salinity). The output result of the analysis will be displayed as a feature class and the clustering will be highlighted. The new output feature class will have the following fields in the attribute table: Local Moran's I index (LMiIndex), z-score (LMiZScore), pseudo p-value (LMiPValue), cluster/outlier type (COType), in addition to other fields from the original layer. The z-scores and p-values are measures of statistical significance which tell users whether to accept or reject the null hypothesis. The interpretation of the result will be based on the following fields in the attribute table:

A high positive z-score in the attribute table indicates that the surrounding features have similar values (either deep wells or shallow wells).

The COType field will be HH for a statistically significant cluster of high values (deep wells) and LL for a statistically significant cluster of low values (shallow wells).

A low negative z-score (less than −1.4) for a well indicates a statistically significant spatial data outlier. The COType field indicate if the well has a deep well and is surrounded by wells with shallow depth (HL) or if the well has a shallow depth and is surrounded by wells with deep wells (LH).

No permutations are used to determine how likely it would be to find the actual spatial distribution of the wells you are analyzing. For each permutation, the neighborhood values around each feature are randomly rearranged and the Local

Moran's I value calculated. The result is a reference distribution of values that is then compared to the actual observed Moran's I to determine the probability that the observed value could be found in the random distribution. The default is 499 permutations; however, the random sample distribution is improved with increasing permutations, which improves the precision of the pseudo p-value.

Scenario 6: In Amman-Zarqa basin, there are many groundwater wells drilled for agricultural development and they are tapping two aquifer systems: the carbonate and basalt aquifers. The wells that penetrate the basalt are in general deeper than the wells tapping the carbonate aquifer, especially in the eastern part of the basin. Your duty is to identify if there is a clustering based on the wells' depth.

GIS Approach
1. Insert Data Frame and call it **Clustering**
2. Integrate the Well.shp, Geology.shp, and WWTP.shp from \\Ch15\Data\Q6 folder
3. R-click Well/Properties/Definition Query/click Query Builder
4. Type the following SQL statement "Well_Depth" > 0, then click OK/OK

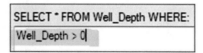

SELECT * FROM Well_Depth WHERE:

Well_Depth > 0

Result: the well number decrease from 2039 to 1787.

5. D-click Geology layer/Symbology/Categories/Unique values, Value Field = Lithology/Add All Values/Color Ramp = Pastels/OK

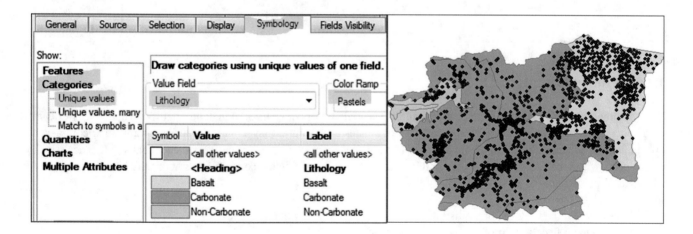

Run Cluster and Outlier Analysis (Anselin Local Moran I)
This method allows you to use a distance of your choice in order to find a significant number of neighbors. A 1000 m and Euclidean distance will be used, wells outside the 1000 m for a target well are ignored in the analyses for that well.

6. ArcToolbox/Spatial Statistics Tools/Mapping Clusters
7. D-click "Cluster and Outlier Analysis (Anselin Local Moran I)"
8. Input Feature Class: Well
9. Input Field: Well_Depth
10. Output Feature Class: \\Result\MICluster1000.shp
11. Conceptualization of spatial relationship: FIXED_DISTANCE_BAND
12. Distance Method: EUCLIDEAN DISTANCE

13. Distance Band: 1000
14. Number of Permutations: 999
15. OK

Result: The MICluster1000 layer will be displayed in the TOC with five classes.

16. R_click the symbol of "**Not significant**" and change its color to **Grey 10%**, the "**High-High Cluster**" to **Ginger Pink**, "**High-Low Outlier**" to **Leaf Green**, "**Low-High Outlier**" to **Solar Yellow**, and the "**Low-Low Cluster**" to **Cretan Blue.**

Interpretation: 192 wells have HH records, which means statistically significant cluster of deep wells and surrounding by deep wells. The depth of the wells in meters range from 230 to 675.

387 wells have LL records, which means statistically significant cluster of shallow wells and surrounding by shallow wells. The depth of the wells in meters range from 5 to 217.

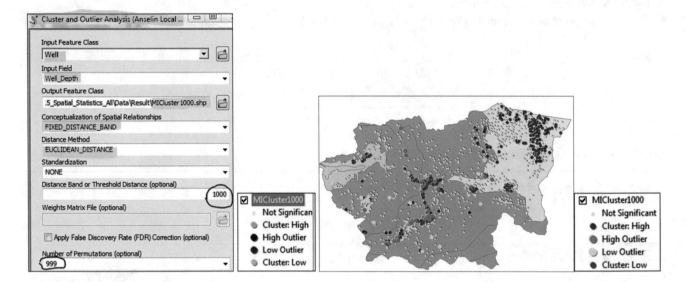

Hot Spot Analysis (Getis-Ord GI*)

This is another method to identify statistically significant spatial clusters of wells of high depth (hot spots) and Shallow depth (cold spots) using the Getis-Ord Gi* statistic. The tool creates a new output layer with a z-score, p-value, and confidence level bin (Gi_Bin) for each well in the input layer. The z-scores and p-values are measures of statistical significance which tell the users whether to accept or reject the null hypothesis. The Gi_Bin field also identifies statistically significant hot and cold spots as below:

- Wells (+3 bins) reflect "Hot Spot" statistical significance with a 99% confidence level
- Wells (+2 bins) reflect "Hot Spot" statistical significance with a 95% confidence level
- Wells (+1 bins) reflect "Hot Spot" statistical significance with a 90% confidence level
- Wells (−3 bins) reflect "Cold Spot" statistical significance with a 99% confidence level
- Wells (−2 bins) reflect "Cold Spot" statistical significance with a 95% confidence level
- Wells (−1 bins) reflect "Cold Spot" statistical significance with a 90% confidence level
- Well with 0 bin indicates no apparent spatial clustering

A high z-score and small p-value for a well indicates a spatial clustering of deep well, while a low negative z-score and small p-value indicates a spatial clustering of shallow wells. The higher (or lower) the z-score, the more intense the clustering. A z-score near zero and with 0 bin indicates no clustering.

Scenario 7: Your advisor asked you to use the wells from the previous scenario to identify the hot and cold spot in Amman-Zarqa basin

1. Insert Data Frame and call it **Hot and Cold Spot**
2. Integrate the Well.shp and Geology.shp, from **Clustering** data frame

3. ArcToolbox/Spatial Statistics Tools/Mapping Clusters
4. D-click "Hot Spot Analysis (Getis-Ord Gi*)"
5. Input Feature Class: Well
6. Input Field: Well_Depth
7. Output Feature Class: \\Result\HotSpot.shp
8. Conceptualization of spatial relationship: FIXED_DISTANCE_BAND
9. Distance Method: EUCLIDEAN DISTANCE
10. Distance Band: 1000
11. OK

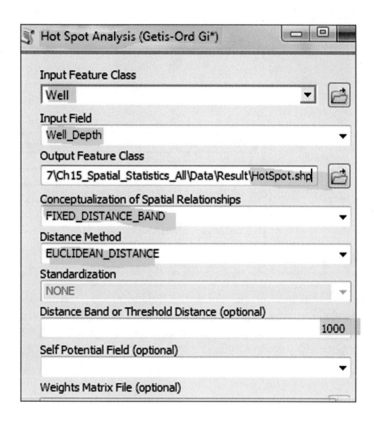

Result: The HotSpot layer will be displayed in the TOC with seven classes.

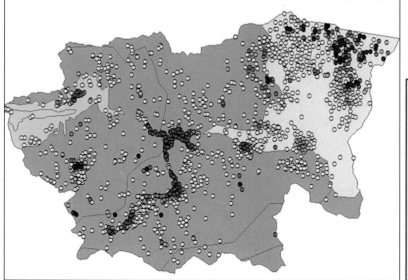

12. Open the attribute table of HotSpot Layer
13. R-click the Gi_Bin/Summarize
14. Select a field to summarize: Gi_Bin
15. Well_Depth: check Average
16. GiZScore: check Average
17. Specify output table: HotCold.dbf
18. OK and open the HotCold.dbf table

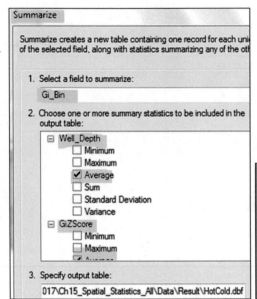

OID	Gi_Bin	No of Wells	Average Well Depth	Average ZScore
0	-3	305	82.74	-4.05
1	-2	85	115.66	-2.27
2	-1	76	119.34	-1.79
3	0	1064	245.04	0
4	1	65	380.42	1.79
5	2	103	430.07	2.3
6	3	89	459.55	3

Interpretation: The HotCold.dbf table includes seven records: Cold Spot, Hot Spot, and no clustering. The cold and hot spot each consists of three groups. A cold spot is depicted by a 99%−90% confidence interval which shows that they are shallow wells and have a negative average Gi_Bin and ZScore values. A Hot Spot is depicted by a 99%−90% confidence interval which shows that they are deep wells with a positive average ZScore and Gi_Bin values. Wells with ZScore close to 0 and 0 bin reflect no clustering.

Proximity and Network Analysis

Introduction

Proximity analysis is an important function in GIS as it covers a wide range of topics that help in answering many spatial questions, such as

1. How close is the observation well to a treatment plant?
2. Do any wells fall within 500 m of a fault system?
3. What are the distances between the wells and the treatment plant?
4. What is the nearest or farthest well from the dam?
5. What is the shortest street network route from the water tower reservoir to the towns?

Proximity tools can be applied in the vector and raster format. The vector-based tools vary in the types of output they produce and can be explained briefly in this chapter.

Proximity Analysis in Vector Format

Buffer analysis is used for identifying areas surrounding any type of feature, whether it is point, line or polygon. The buffer polygon is created to a specified distance around an input feature. The output polygon features can be used as an input to overlay tools (union, intersect, erase, and spatial join). **Multi-Ring Buffer** creates a new feature class of buffer features using a set of buffer distances. Buffer function doesn't take into consideration any physical obstacle that might exist in the area of buffering.

Select by Location: After generating the buffer, user can use the **select by location** using different relationships between the buffer and the source feature under investigation. The select by location does not draw a boundary, but select the features that are determined by the relationship between the source and target layers.

Near function is selecting one feature of a set and then calculating the distance to all other features in the same set. The Near tool adds a new field called "distance" in the attribute table of one of the input layers. The distance will be calculated based on the map unit of the coordinate system of the map document.

Point Distance is similar to the Near function, but it generates an independent table. It calculates the distance from each point in one layer to all of the points within a given radius in another layer. The generated table can be used for further statistical analyses.

Electronic Supplementary Material: The online version of this chapter (https://doi.org/10.1007/978-3-319-61158-7_16) contains supplementary material, which is available to authorized users.

W. Bajjali, *ArcGIS for Environmental and Water Issues*, Springer Textbooks in Earth Sciences, Geography and Environment, https://doi.org/10.1007/978-3-319-61158-7_16

Desire Lines (Spider diagram) will draw a line from each record to the one selected feature to identify the exact location. The Desire lines tool is part of the business analyst and shows which customers visit which stores. A line is drawn from each customer point to its associated store point, making it easy to see the actual area of influence of each store. This tool can be used in environmental related problems.

Scenario 1: The region of Dhuleil-Samra, in Jordan, is considered an arid area and groundwater, which is scarce and has low water quality, is the only source for domestic use. As a hydrogeologist working for Water Authority you have been asked to explore the possibilities of finding two wells with good water quality. One well in each region: Dhuleil and Samra. The selected two wells would be used for water supply in the two regions. The two wells should have the following criteria:

1. The well in the Samra region should be 2.5 km away from the Wastewater Treatment Plant (WWTP) and serve only the towns in the Samra region
2. The well in Dhuleil region should be 2.5 km away from the stream and serve only the towns in the Dhuleil region
3. Both selected wells should have total dissolved solids (TDS) and nitrate (NO_3) less than 1000 mg/L and 20 mg/L respectively

After identifying the two suitable wells, a pipeline will be extended from each selected well to each town in the designated region. In your final report you decided to provide a table and map showing the distance from each selected well to the towns in both regions.

Scenario 2: Your boss asked you to work with another scenario, by choosing the "**Hay Arnous**" town and build, in it, a big Water Supply Tower (WST). The water from the two selected wells will be pumped into the WST and then distribute it to the whole regions by gravity. To consider this approach you need to know the number of towns that will be served and their population. To carry the work you should do the following:

1. The **Hay Arnous** town should be buffered into three rings with a radius of 4, 11, and 18 km
2. Find how many towns are located in each ring and their total population

Ring radius (km)	No. of towns	No. of populations
4		
11		
18		

Scenario 3: You are working in Dhuleil as a hydrogeologist and you have been given an assignment by your supervisor to verify the argument that the dam in the north-east of the Dhuleil region is playing a role as an artificial recharge. Therefore, the water stored behind the dam in the rainy season will infiltrate into the subsurface aquifer and improve its water quality. The water quality will be checked through two parameters, TDS and NO_3, in the wells that are located within 10 km radius from the dam. Low concentration of TDS and NO_3 in the wells close to the dam means the argument is true, otherwise it is false.

GIS Approach to Solve Scenario 1

1. Start ArcMap and integrate the Region, Road, Stream, Town, Well, and WWTP from \\Ch16\Data01\Q1
2. Click the town symbol select Square 2, Size 7, and the Color = ginger pink/OK
3. Click the symbol of the stream and select the river symbol in the Symbol Selector
4. D-click Region layer/click Symbology tab/Categories, Unique values, Value Field = Name, Uncheck all other values, Click Add All Values, click OK

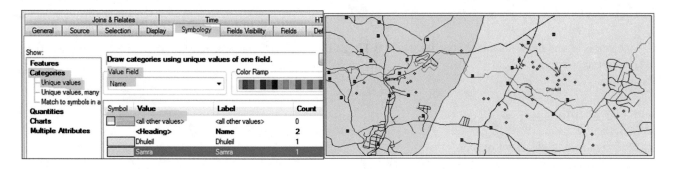

5. In the TOC, r-click Region layer/Label Features

   Question: How do you make the font of the label Time New Roman, size 14, and place it on the top of the two regions?

Buffer the WWTP in the Samra Region

6. Click ArcToolbox, r-click empty place/click Environment, then Workspace
7. Current Workspace: \\Data01\Q1
8. Scratch Workspace: \\Result
9. OK

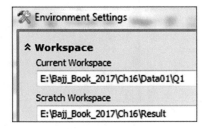

10. ArcToolbox/Analysis Tools/Proximity/d-click Buffer
11. Input Features: WWTP
12. Output Feature Class: \\Result\WWTP_Buffer
13. Linear units: 2.5 km
14. Accept the other default
15. OK

16. R-click WWTP_Buffer/Properties/Display tab/Transparent 50%

Next step is to find the wells located outside the **WWTP_Buffer** zone in Samra that have a TDS and NO$_3$ less than 1000 and 20 mg/L respectively.

Select By Location

17. Selection menu/Select By Location
  a. Selection Method: Select features from
  b. Target layer(s): Well
  c. Source layer: WWTP_Buffer
  d. Spatial selection method for target layer: are completely within the source layer feature

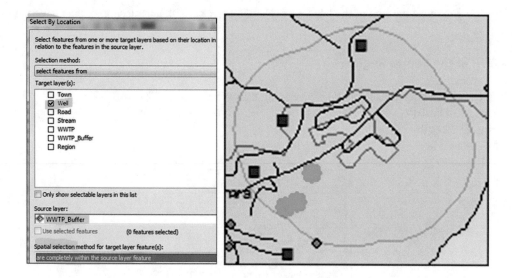

Result: Three wells have been selected inside the buffer zone of the WWTP.

SQL Statement

Now we have to find one well that is located outside the buffer zone in the Samra region and has a TDS and NO$_3$ less than 1000 and 20 mg/L respectively.

18. Open the attribute table of the **Well.shp**

19. Click switch selection button

20. Click Table Option drop-down arrow/Select by Attribute/fill the Dialog Box as follow
21. Method: Select from current selection
22. Write the SQL Statement:

"Region" <> 'Dhuleil' AND "TDS" < 1000 AND "NO3" < 20

23. Apply/Close

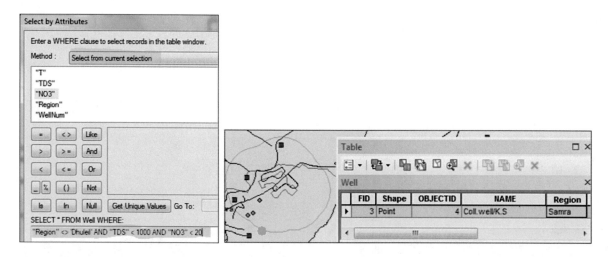

Result: One well has been selected in the Samra region.

In order to proceed you have to add a new field in the attribute table of the **Well**

24. Make sure the attribute table of the **Well.shp** is open
25. Table Option/Add Field/Name it "**Suitable**"/Type = Text/Length = 12
26. R-click "Suitable" field/Field Calculator/click Yes/
27. In the Field Calculator type "Yes" under Suitable, then click OK

28. In the Well attribute table click Switch Selection button
29. R-click "Suitable" field/Field Calculator/click YES/
30. In the Field Calculator type "No" under Suitable, then click OK

Result: The attribute table of the well that has a TDS and NO_3 less than 1000 and 20 mg/L populated with "YES" and the rest with "NO" in the "Suitable" field.

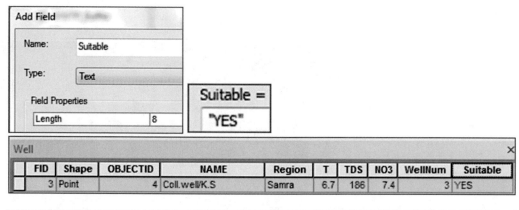

Buffer the Stream in the Region

31. ArcToolbox/Analysis Tools/Proximity/d-click Buffer
32. Input Features: Stream
33. Output Feature Class: \\Result\Stream_Buffer
34. Linear units: 2.5 km
35. Accept the other default
36. OK
37. Make the Stream_Buffer 50% transparent

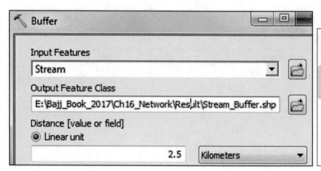 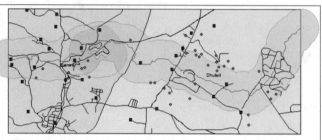

38. Selection menu/Select By Location
    a. Selection Method: Select features from
    b. Target layer(s): Well
    c. Source layer: Stream_Buffer
    d. Spatial selection method for target layer: "are completely within the source layer feature"
39. OK

Result: 20 wells from both regions have been selected inside the stream buffer.

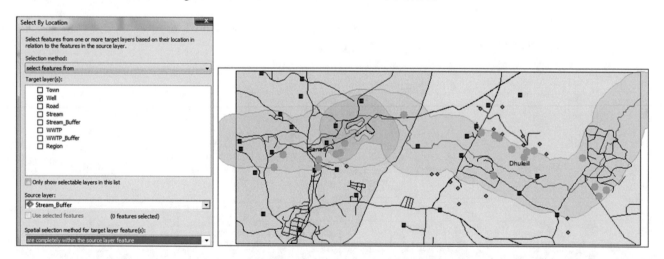

Now, we will find the wells located outside the Stream Buffer zone in Dhuleil area that have TDS and NO_3 less than 1000 and 20 mg/L respectively

40. Open the attribute table of the **Well** layer

41. Click the Switch Selection ⬚ button
42. Table Option/Select By Attribute/
43. Method: Select From Current Selection

44. Write the following SQL statement

"Region" <> 'Samra' AND "TDS" < 1000 AND "NO3" < 20

45. Apply/Close

Result: Two wells have been selected outside the buffer, therefore, you are going to choose the well that has lower TDS.

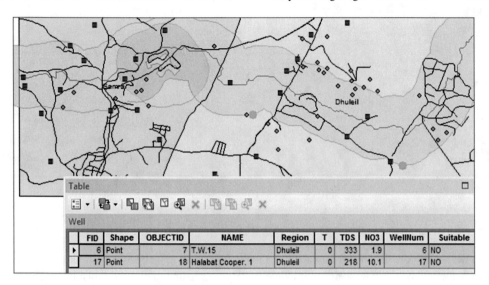

46. In the attribute table of the Well layer click the first cell of the record that has a TDS = 333, the color of the record will turn to yellow color

	FID	Shape	OBJECTID	NAME	Region	T	TDS	NO3	WellNum	Suitable
	6	Point	7	T.W.15	Dhuleil	0	333	1.9	6	NO
▶	17	Point	18	Halabat Cooper. 1	Dhuleil	0	218	10.1	17	NO

47. In the attribute table click the Unselect Highlighted button

Result: The well that has TDS 333 will be deselected, and well that has OBJECTID =18 remains selected.

	FID	Shape	OBJECTID	NAME	Region	T	TDS	NO3	WellNum	Suitable
	17	Point	18	Halabat Cooper. 1	Dhuleil	0	218	10.1	17	NO

48. R-click Suitable field/Field Calculator/type "YES" under Suitable, then click OK

	FID	Shape	OBJECTID	NAME	Region	T	TDS	NO3	WellNum	Suitable
	17	Point	18	Halabat Cooper. 1	Dhuleil	0	218	10.1	17	YES

Definition Query

To proceed with the analysis we have to hide all the wells and keep only the two selected wells in the Samra and Dhuleil regions.

49. R-click Well/Properties/Definition Query tab/click Query Builder/
50. Type the following SQL statement "Suitable" = 'YES'
51. OK/OK
52. Click the symbol of the Well layer, use Circle 2, Size 10, and color = Ultra Blue

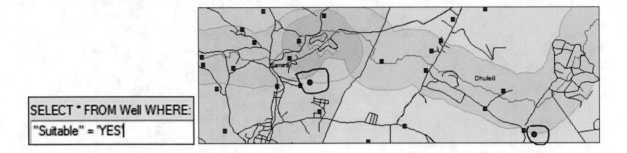

Near Tool

The Near function determines the distance from each town to the two selected nearest wells in the study area. The Near tool will not generate any output layer, but will add two new fields to the attribute table of the Town layer. The two fields that are added will be called NEAR_FID and NEAR_DIST. The NEAR_FID contain the feature ID (WellNum) of the Well layer. The NEAR_DIST stores the distance from each well to the nearest town. The value of this field is in meters because the coordinate system of both data is in Palestine_1923_Palestine_Belt (custom UTM).

1. Insert New Data Frame and call it Near Proximity
2. Drag the Region, Town, and Well layers from the Dhuleil-Samra data frame into Near Proximity data frame

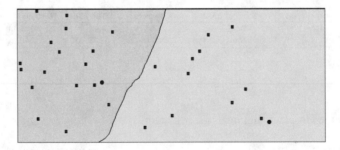

3. ArcToolbox/Analysis Tools/Proximity/d-click Near
   a. Input Features: Town
   b. Near Features: Well
   c. Accept the rest as a default
4. OK
5. Open the attribute table of the Town (2-fields are added NEAR_FID and NEAR_DIST)

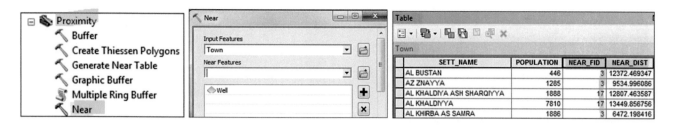

Result Explanation: In the attribute table of the town two fields were created the **Near_FID** and **Near_Dist**. The **Near_FID** has two variables (3 and 17). Well No 3 is the **WellNum** in the well layer located in Samra area. Well No. 17 is the well that is located in Dhuleil area. The **NEAR_DIST** field shows the distance between each well and the towns in each region of both Samra and Dhuleil. For example, the distance between well No 17 and Al Mabruka town is 12.65 km, while the distance between well No. 3 and Mabruka town is 13.29 km. Therefore, the Near distance will associate Mabruka town with well No. 17 because the distance is shorter.

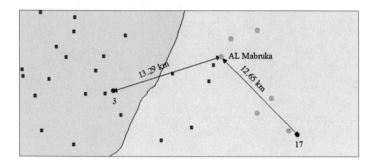

Desire Lines (Spider Diagram) Tool

Desire Lines will draw lines from each well to the closest towns in the two regions. Spider diagram can be performed using the "**Desire Lines**" tool, which is part of the Business Analyst. The tool can be used in environmental related problems to see, for example, the influence of a landfill on the groundwater observation wells. Therefore, a line is drawn from the landfill point to its nearest wells, making it easy to see the actual area of influence of the landfill.

Note: To run this function make sure you have the Business Analyst Standard installed and the extension is checked.

6. Insert New Data Frame and call it Spider Diagram
7. Drag the Region, Town, and Well layers from the **Near Proximity** data frame into Spider Diagram data frame
8. ArcToolbox/Business Analyst Tools/Analysis/d-click Desire Lines tool
    a. Store Layer: Town
    b. Store ID Field: Near_FID
    c. Store to Use: ALL
    d. Customer Layer: Well
    e. Customer Layer: WellNum
    f. Accept the default
    g. Output Table: \\SpiderDiagram.shp
9. OK

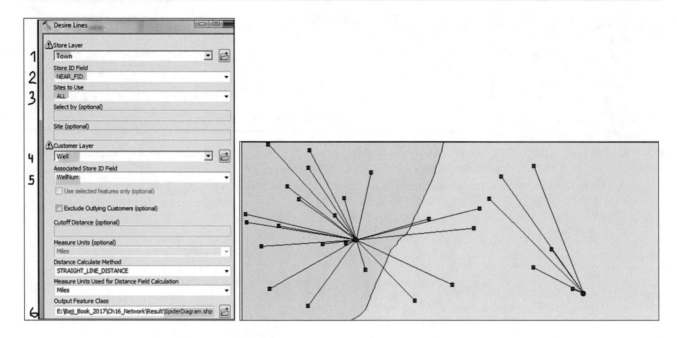

Result: A line will be drawn from each well to the closest town.

GIS Approach to Solve Scenario 2

Use the Multi-Ring Buffer Around the Hay Arnous Town

1. Insert Data Frame and call it Water Reservoir
2. Drag from "Spider Diagram" data frame the region, Town and Well layers into Water Reservoir data frame
3. Open the attribute table of the Town/click Table Options/Select By Attribute
4. Write the SQL statement: **"SETT_NAME" = 'HAY ARNOUS'**
5. Apply/Close

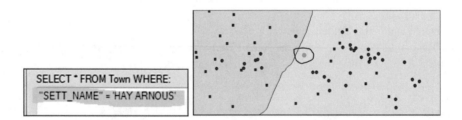

Result: The 'HAY ARNOUS' town is selected and is located in the middle of study area.

 6. ArcToolbox/Analysis Tools/Proximity/d-click Multiple Ring Buffer
 7. Input Features: Town
 8. Output Feature Class: Arnous_Buffer
 9. Distance: type 4 click the + sign, Type 11 click the + sign, Type 18 click the + sign
10. Buffer Unit: kilometer
11. Dissolve Option: All
12. OK

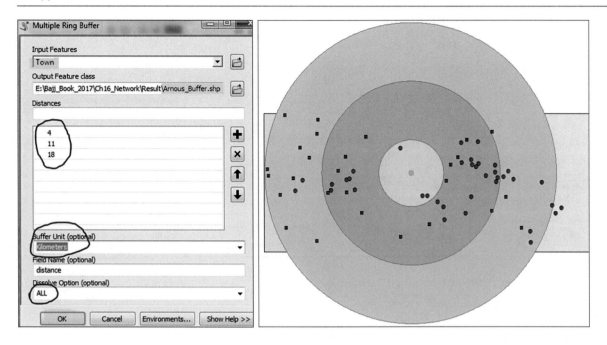

Result: A 3 ring buffer is created around the town of 'HAY ARNOUS'.

13. Open the attribute table of the Arnous_Buffer/select the third record (18 km distance)

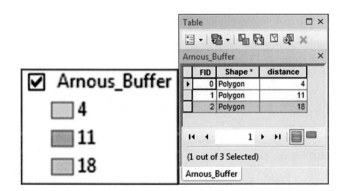

14. Selection/Select By Location/
    a. Selection method: select feature from
    b. Target layer(s): Town
    c. Source Layer: Arnous_Buffer
    d. Check "Use selected features"
    e. Spatial selection within the source layer feature(s): "are completely within the source layer feature"
15. OK

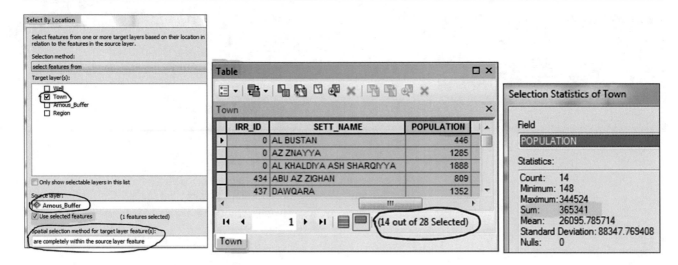

16. Open the attribute table of the Town

   Result: You will see that there are 14 towns selected.

17. R-click the field Population/Statistics

   Result: Total Population is 365,341.

18. Repeat the previous steps and write down number of towns and the total population of each Buffer zone in the table below.

Ring radius (km)	No. of towns	No. of populations
4		
11		
18	14	365,341

GIS Approach to Solve Scenario 3

Proximity Analysis in Raster Format

In this section **Euclidean Distance** applications, which are raster based, will be applied. In addition, the vector based application of the **Point Distance** will also be applied to evaluate the effect of the dam on the quality of groundwater.

Euclidean Distance: The raster-based Euclidean distance tool measures distances from the center of source cells to the center of destination cells. The Spatial Analyst extension can perform analysis where the output layer is in raster format. One of the analysis that can be applied in Earth sciences is the distance surface. This method create a continuous layer from a vector input layer. The vector layer can be point, line, or a polygon such as groundwater well, stream, or a treatment plant.

Point Distance Method

The point distance determines the distances from input point features to all points in the near features within a specified search radius. The tool is similar to the Near tool, but creates a table with distances between the two sets of point layers. If the default search radius is used, distances from all input points to all near points are calculated.

GIS Approach to Solve Scenario 3

1. Insert Data Frame and call it "**Recharge**"
2. Integrate the **Dam**, **Region**, **Stream**, and **Well** from \\Data01\Q2 folder
3. Click the Dam symbol, type "Dam" in the search window in the Symbol Selector dialog box, and click Search button
4. Click the Dam symbol, then click OK
5. Click the well symbol, select Circle 2, color = blue, and size = 9, then click OK
6. Click the stream symbol, select River symbol, then click OK
7. R-click the Region symbol, change the color to Hollow

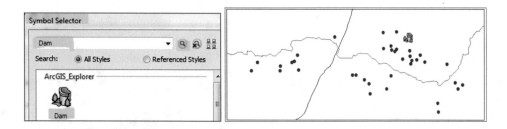

8. At the bottom of ArcToolbox/r-click an empty place
9. Click Environments/Click Workspace
10. Current Workspace: \\Ch16\Data01\Q2
11. Scratch Workspace: \\Ch16\Result
12. Click Processing Extent
13. Extent: Same as layer Region
14. Click OK
15. ArcToolbox/Spatial Analysis Tools/Distance/d-click Euclidian Distance
16. Input raster or feature source data: Dam
17. Output distance raster: \\Result\Dam_Distance.tif
18. Maximum distance: 10000
19. Output cell size: 100
20. OK

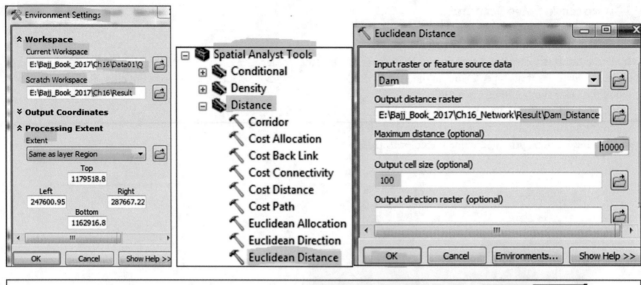

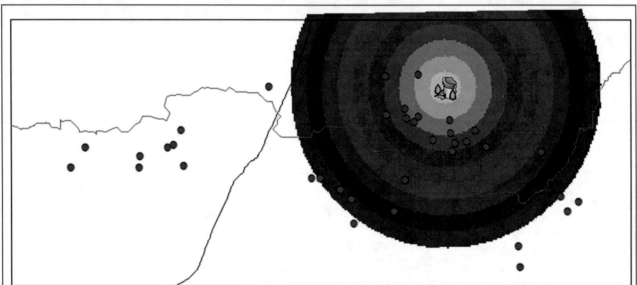

Result: The output **Dam_Distance** raster is classified into 10 classes 1000, 2000, 3000, 4000, 5000, 6000, 7000, 8000, 9000, and 10,000.

Point Distance Tool

10. ArcToolbox/Analysis Tools/Proximity/d-click Point Distance tool
    a. Input Features: Dam
    b. Near Features: Well
    c. Output Table: \\Dam_Well.dbf (save as .dbf)
    d. Search Radius: 10000 m
11. OK

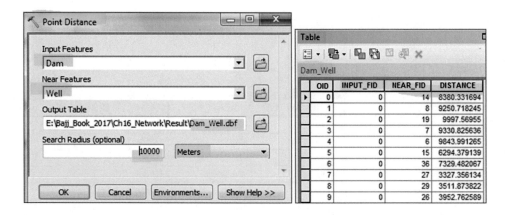

Result: The output is a dbf table.

- In the TOC open **Dam_Well.dbf table**, you will see "DISTANCE" field that show the distance between the dam and the wells located within 10000 m from it
- Input_FID: The Dam feature ID, which has 0 value
- NEAR_FID: The feature ID of all Wells located within 10000 m from the dam

Next step is to join the **Dam_Well.dbf** with the **Well** layer based on a common field. The **WellNum** field in the **Well** layer is identical to the **NEAR_FID** field in the **Dam_Well**.

12. R-click Well/Join and Relates/click Join
13. What do you want to join to this layer? **Join attributes from a table**
14. Choose the field in this layer that the join will be based on: **WellNum**
15. Choose the table to join to this layer: **Dam_Well**
16. Choose the field in the table to base the join on: **NEAR_FID**
17. Join Options: **Keep all records**
18. Click Validate Join
19. Click Yes to create an index for the join field
20. Click Close
21. Ok

Result: The **Dam_Well.dbf** is joined to the **Well** layer.

22. Open the attribute table of the Well layer

The Distance field in the Well layer is populated with zero <NULL> and number larger than zero. The zeros are the wells that are located more than 10,000 m from the dam. The wells that are located within 10,000 m from the dam, their distance from the dam are listed in the Distance field.

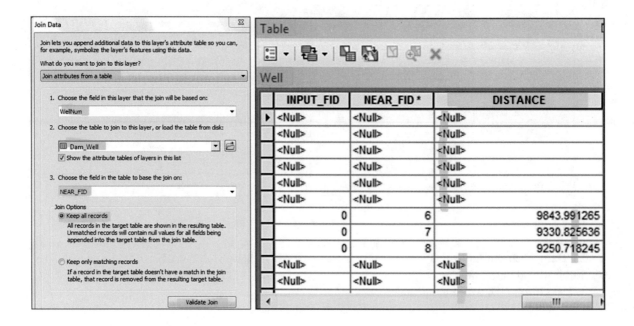

Definition Query

23. R-click Well layer/Properties/Definition Query/click Query Builder
24. Type SQL statement "Dam_Well.DISTANCE" > 0
25. OK/OK

Result: All the wells located farther than 10,000 m will disappear.

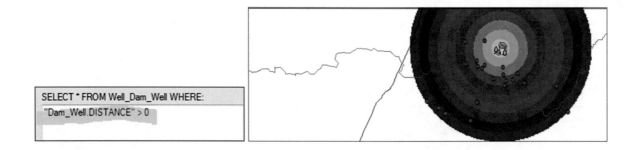

Well Classification

To verify the effect of the dam, the TDS (salinity) and nitrate concentration of the wells will be classified.

Salinity (TDS) Classification
26. D-click Well Layer/click the Symbology tab/click Quantities/click Graduate symbols
27. Field Value: TDS, click Classify/Method Manual/Under Break Value
28. Type, 500, 1000, 2000, 3000, and keep the last value 4160/click an empty place
29. Click OK/click Label Header/check Show thousands separator/OK
30. Change the highest TDS symbol to red, pink, green, blue, then cyan/OK

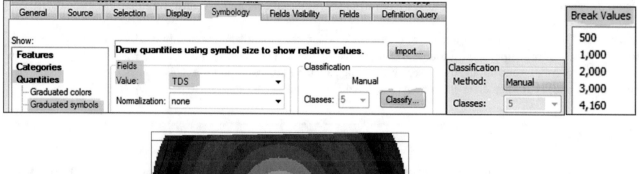

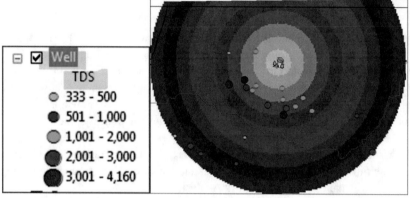

Result: The closest wells to the dam are the ones that have salinity between 1000 and 2000 mg/L. Wells that are located further away from the dam have higher salinity.

Nitrate (NO₃) Classification

31. D-click Well Layer/click the Symbology tab/click Quantities/click Graduate symbols
32. Field Value: NO₃/click Classify/Method Manual/Under Break Value
33. Type, 20, 45, 70, 100, and keep the last value 145/click an empty place
34. Click OK/click Label header/click Format Label/Number of decimal places = 0
35. Click OK
36. Change the NO₃ symbol (highest to lowest) to red, pink, green, blue, then cyan/OK
37. OK
38. File Save as \\Result\Recharge.mxd
39. Exit your ArcMap

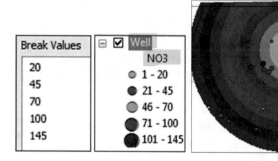

Result: The lowest concentration of nitrate are the wells that are located in close proximity to the dam.

Conclusion: The wells are located in an agricultural area, where the groundwater is mainly used for irrigation. The high salinity and nitrate concentration are attributed mainly to return flow and application of chemical fertilizers. Despite the high concentration of salinity and nitrate, the closest wells to the dam have less concentration. This indicates that the stored water behind the dam leaks downgradient and improves the quality of water.

Network Analyst

This section will perform different GIS functions in environmental related problems using the Network Analyst. The Network Analyst is a powerful tool in a way that it is different than the proximity analysis performed in the previous section. The applications will overcome the concept of the straight distance related to some application such as buffering and Euclidian distance calculation. The network analyst will overcome any natural barrier such as hills, lakes, or where there is absolutely no network of a street system. The network analyst will use the actual distance that is associated with the street feature, which is an important feature in the application. This approach is more accurate than using the near function or spider diagram model.

To use the Network Analyst you need to have a line feature that has connectivity such as a street, pipe, railroad, etc. If, for example, you have a street, you need to calculate the time from the length of the street and the speed of each segment of the street (Distance = speed * time). The time will be used as the cost in the Network Analyst.

You will perform two scenarios in the Dhuleil-Samra region related to the water supply problem. The data includes a street feature class that has a Shape_Length and Speed field in its attribute table. A new field "MINUTES" will be added to the Street feature class, which will allow network routing.

Scenario 1: Samra - Dhuleil region has a shortage of water supply in summer time. The Water Authority decided to use two good quality water wells to supply the towns in the region with potable water using water trucks. Your duty is to find how long time it requires the truck to supply the towns in the study region with water.

Scenario 2: After finding the time required to cover the towns in the region, you want to find the actual path and time that the water truck will take from each individual well to each town.

Your duty to solve scenario 1 by doing the following:

1. Create Network Dataset (SupplyTravelTime)
2. Build NetworkDatset in Network Analyst
3. Run the "Add Location Function"
4. Calculate the True Path and Total Time between Wells and Towns

GIS Approach

1. Launch ArcMap and rename the Layer Data Frame and call it **Service Area**
2. Integrate Region, Town, and Well_Supply feature classes from \\Data02\Q4\Region.mdb
3. Classify the Region based on the Name field
4. Symbolize the Town layer (Square 2, size 7, and ginger pink color)
5. Symbolize the Well_Supply layer (Circle 2, size 10, and blue color)

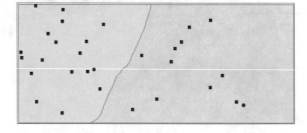

6. Open Catalog window/r-click Result/New/Personal Geodatabase
7. Rename the geodatabase **"Water Supply.mdb"**
8. R-click Water Supply.mdb/New/Feature Dataset/call it **SamraDhuleil**
9. Click Next

10. Click the Add Coordinate System button drop-down arrow

11. Click Import/browse to \\Q4\Region.mdb\click Well_Supply/Add/Next/Next/Finish

Note: The coordinate of the Well_Supply feature class is Palestine_1923_Palestine_Belt and this coordinate system is now assigned to the **SamraDhuleil** Feature Dataset.

12. R-click SamraDhuleil Feature dataset/Import/Feature Class (single)
13. Input Features: \\Data02\Q4\Street.shp
14. Output Location: \\Result\Water Supply.mdb\SamraDhuleil
15. Output Feature Class: Street
16. OK

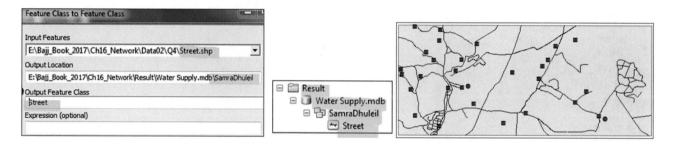

Result: The **Street** feature class will be added to the TOC.

17. Open the attribute table of the Street feature class, see that the table has two fields "Shape_Length" and "Speed"

   Note: The Speed is in kilometer per hour, and the Shape_Length is in meter.
   Next step is to add a new field "**MINUTES**" to the **Street** feature class in order to calculate the time from the "Shape_Length" (length of the street) and "Speed" fields. The "**MINUTES**" field allows user to perform network routing.

18. R-click street feature classes in the Catalog window/Properties/click Field tab
19. Under **Field Name** write "MINUTES" Data Type "Double"
20. Click OK

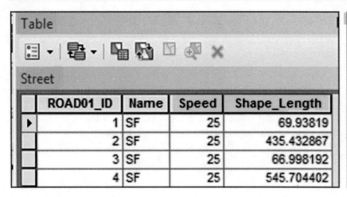

21. In TOC/R-click MINUTES/Field Calculator/click "Yes"
22. In the Field Calculator dialog box, type the following Statement:

$$\text{MINUTES} = \textbf{[Shape\_Length]/(([Speed] * 1000)/60)}$$

23. Click OK

    Result: The Minutes field is now calculated.

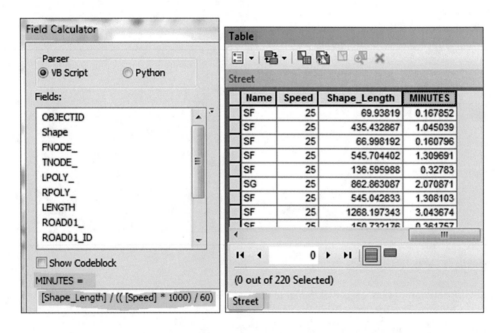

Create Network Dataset

24. In Catalog window/r-click **SamraDhuleil** Feature Dataset in **WaterSupply.mdb**/click New/Network Dataset
25. Name: **SupplyTravelTime** (choose 10.1 version) and click Next
26. Check the Street and click Next
27. Do you want to model turns in this network? Check "Yes" and click Next
28. Accept the default "Connectivity" and click Next
29. How would you like to model the elevation of your network features? Check "Using Elevation Field" and click Next
30. Specify the attributes for the network dataset, accept the default and click Next
31. Accept the default/click Next
32. Do you want to establish driving setting for this network dataset? check Yes, and click Next
33. Click Finish
34. A dialog box will display "The new network dataset has been created. Would you like to build it now?" Click No
35. Click Yes to add all feature classes that participate in "SupplyTravelTime" to the map

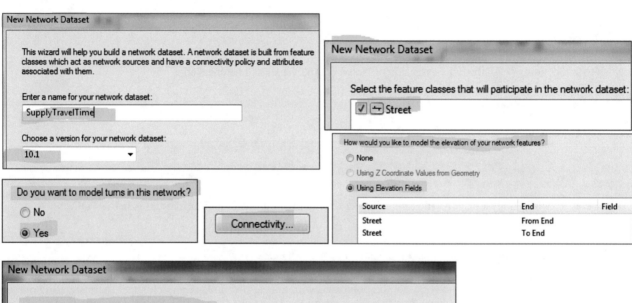

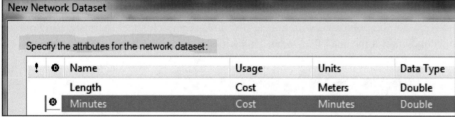

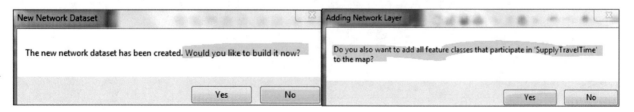

Result: The new network dataset SupplyTravelTime is created and SupplyTravelTime_Junction, Street, and SupplyTravelTime feature classes are added to the TOC.

Note: If you open the attribute table of the "SupplyTravelTime_Junctions", you will find that it is empty.

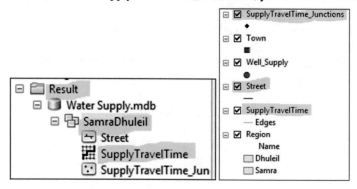

Build the Network

To work with the Network Analyst, the extension of Network Analyst should be checked.

36. Click Customize menu/Toolbar/check Network Analyst

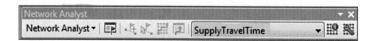

Note: The Network Analyst displays the SupplyTravelTime network as the default because it is in the TOC.

37. Click the Build Network Dataset button on the Network Analyst

Result: The SupplyTravelTime_Junction will be populated with nodes, two nodes for each street segment, one node at the beginning of the segment, and the second node at the end of segment. This will allow you to run several types of network analysis, such as New Service Area and New Closest Facility.

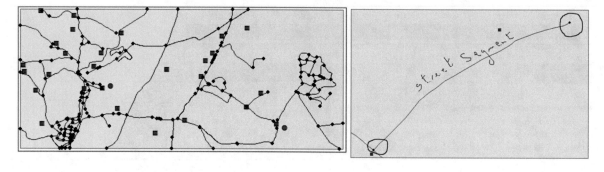

New Service Area

The Network Service Area is a region that covers all accessible streets in the study region. The tool show, for example, within 10 min the water truck will travel from the water supply well along the street network and will include all the streets that can be reached within 10 min time.

Service areas created by Network Analyst also help evaluate accessibility. Concentric service areas show how accessibility varies with impedance. Once service areas are created, you can use them to identify how much land, how many people, or how much of anything else is within the neighborhood or region.

38. From Network Analyst Drop Arrow/Choose New Service Area

Result: Service Area is added to the TOC and has the following feature classes:

1. Facilities
2. Point Barrier
3. Line
4. Line Barrier
5. Polygon
6. Polygon Barrier

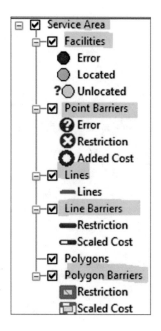

Note: The attribute table of all these feature classes are empty.

Add Locations Tool

This **Add Locations** tool is used to add the **Well_Supply** to the **Network Dataset** so the Well_Supply feature class can be used in the Network Service Area analysis

39. ArcToolbox/Network Analyst Tools/Analysis and d-click Add Locations
    a. Input network Analysis Layer: Service Area
    b. Sub Layer: Facilities
    c. Input Locations: Well_Supply
    d. Name: WellID
    e. Search Tolerance 2000 m
40. Check Snap to Network
41. OK

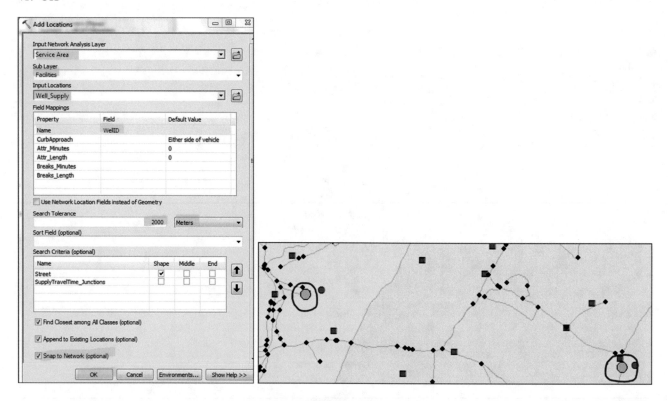

Result: The "Located" that is under the Facilities of the Service Area TOC is now created with two features that represent the two wells (Well_Supply). The located are now snapped to the SupplyTravelTime, which indicate that they have now become part of the Network Dataset and can participate in Network Analysis.

Calculate the Service Area with Certain Travel Time

Ten Minutes Travel Time

The Water Authority now wants to see the areas that will be served with water supply within 10 and then within 30 min by the water truck and to see which would be more sufficient.

42. R-click Service Area/Properties/Analysis Setting tab
    a. Impedance: Minutes (Minutes)
    b. Default Breaks: 10
    c. Direction: check "Away from Facility"
    d. U-Turns at Junctions: Allowed
43. OK

44. On the Network Analyst toolbar click the Solve button ⊞ (4th icon from left)

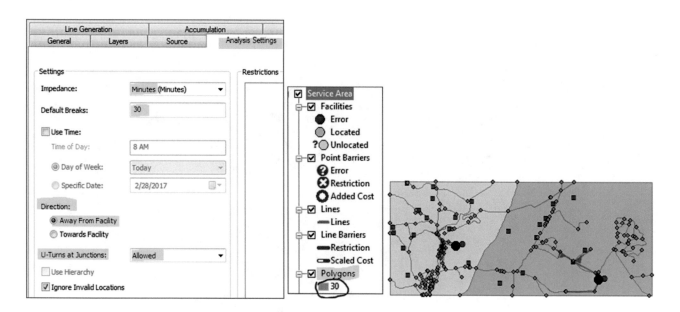

Result: The Polygons under the Service Area have been populated showing the service area that has been covered. The polygon represents the distance along the street network that can be covered in 10-min drive. It is clear that 10 min is not a sufficient time to cover a wider area with water supply by the water truck.

20-Minutes Travel Time

45. R-click Service Area/Properties/Analysis Setting tab
    a. Impedance: Minutes (Minutes)
    b. Default Breaks: 20
    c. Direction: check "Away from Facility"
    d. U-Turns at Junctions: Allowed
46. OK
47. On the Network Analyst toolbar click the Solve button ⊞ (4th icon from left)

Result: 20-minutes is not a sufficient time to cover the whole region with water supply.

Question: Can you run 30-min travel time to see if the water truck can cover the whole region with water supply? Do you think 30-min is sufficient?

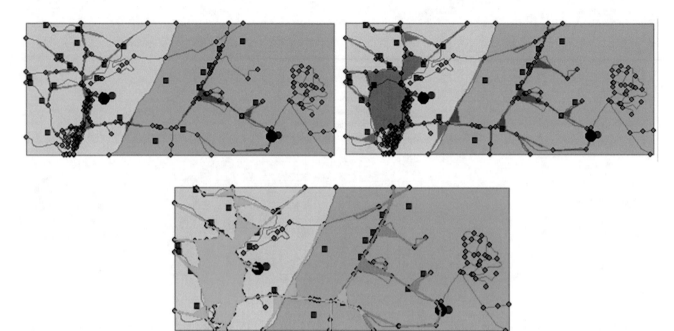

48. R-click the Polygon under Service Area/Data/Export Data
49. Save it in \\Result**ServiceArea_30.shp**
50. "Yes" to add it into the TOC
51. D-Click **ServiceArea_30m**/Symbology/Categories/Unique values/Value Field = Facility ID/Add All Value/change the colors of 1 to blue and 2 to green/OK
52. Save your map in the Result folder \\Result\ServiceArea.mxd

Conclusion: It is clear that a 30-min travel time is sufficient in order to distribute the water from the two wells to all of the communities in the both regions.

Calculate the True Path and Total Time between the Wells and Each Town

Scenario 2: After finding the time required to cover the towns in the region, you want to find the actual path and time that the water truck will take from each individual well to each town. The actual path can be carried out using the "Closest Facility" tool on the Network Analyst.

The "Closest Facility" tool is similar to the "Near" tool that have been used earlier as both measure the distance between two locations. But they are different as the "Near" tool measures the straight line distance, while the "Closest Facility" tool measures the distance along a network.

To solve Scenario 2 you have to do the following:

a) Use the SupplyTravelTime network dataset
b) Calculate the True Path and Total Time between Wells and Towns

GIS Approach

53. Insert Data Frame call it True Path
54. Drag Region, Town, and Well_Supply from Service Area data frame
55. Open Catalog window, drag the **SupplyTravelTime** into the top of TOC from \\Data02\Q5\Water Supply.mdb\ SamraDhuleil
56. Click "YES" to add all feature classes that participate in "SupplyTravelTime"

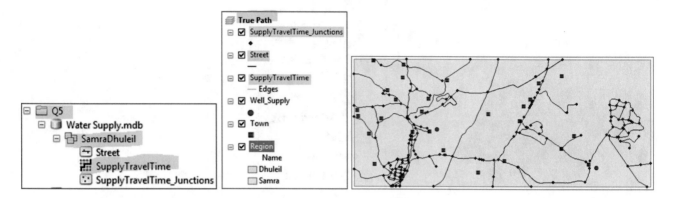

Result: SupplyTravelTime_Junction, Street, and SupplyTravelTime feature classes are now added to the TOC of the True Path data frame.

57. On the Network Analyst toolbar/click the Network Analyst drop-down arrow/click New Closest Facility

   Result: A set of layers are added to the TOC of the Closest Facility layer needed for the analysis and they are:

a) Facilities
b) Incidents
c) Point Barrier
d) Routes
e) Line Barriers
f) Polygon Barriers

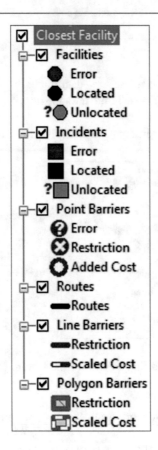

Run Add Locations Tool Between Facilities and Well_Supply

58. ArcToolbox/Network Analyst Tools/Analysis and d-click Add Locations
   f. Input network Analysis Layer: Closest Facility
   g. Sub Layer: Facilities
   h. Input Locations: Well_Supply
   i. Name: WellID
   j. Search Tolerance 2000 m
59. Check Snap to Network
60. OK

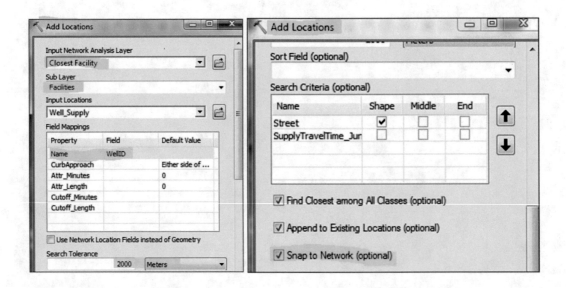

Result: The Located that are under the Facilities of the Closest Facility in the TOC is now created with two features that represent the two wells (Well_Supply). The located are now snapped to the SupplyTravelTime, which indicate that they are now part of the Network Dataset and the can participate in network analysis.

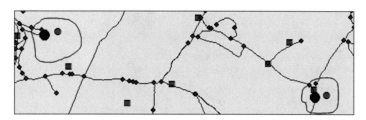

Run Add Location Tool Again

Run "Add Location" tool to relate the town layers to the network

61. ArcToolbox/Network Analyst Tools/Analysis and d-click Add Locations
    k. Input network Analysis Layer: Closest Facility
    l. Sub Layer: Incidents
    m. Input Locations: Town
    n. Search Tolerance 2000 m
62. Check Snap to Network
63. OK

64. On the Network Analyst toolbar click **Solve** button (4th icon from left)

    Result: The Routes path is built and it connects the two wells to each city.

65. D-click Routes feature class under Closest Facility
66. Click Symbology tab/Categories/Unique Value/Value Field = FacilityID
67. Add All Values
68. Change the color of 1 to Fire Red and line width 3, change 2 to green and width 3
69. OK

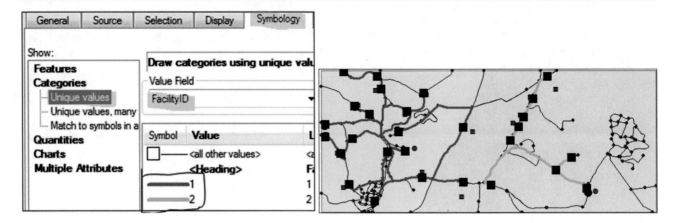

70. Open the attribute table of the Routes
71. You see a field called FacilityID and a field called Total Minutes

Explanation: This shows how much time it takes the truck to travel from each well to each town.

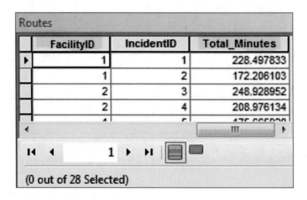

Show Route Direction

The direction button on the Network Analyst tool shows the direction from each well to the towns.

60. Click the direction button 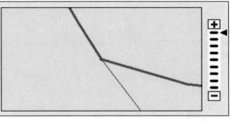 (4ᵗʰ from left)

Result: A Directions (Closest Facility) Dialog Box display shows the route running from the two wells to the towns in the region.

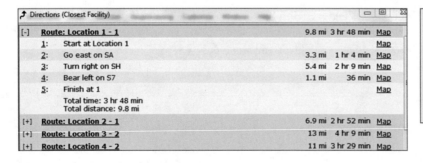

61. Save your map as **Closest Facility.mxd** in the \\Result folder

Introduction

The 3-D Analyst is a software extension that's designed to accomplish different types of analysis such as making the map appear real and making it easy to comprehend. To use the ArcGIS 3D Analyst, it must be installed. The main function of the 3-D Analyst is to create elevation data and display it in three dimensions. The 3-D analyst can also perform slope, aspect, hillshade, surface area, volume calculation and viewshed to determine visibility from any point on a surface.

The 3-D Analyst in ArcMap also allows you to determine line of sight, allows you to create vertical profile, and also digitize 3D features and graphics.

ArcScene allows you to create a raster surface model and Triangulated Irregular Network (TIN) from different elevation input data such as (1) contour line, (2) GPS, and (3) survey points. You can also drape a vector or an image over the TIN or surface models.

The 3-D Analyst adds ArcScene to the ArcGIS application and allows you to integrate the data, visualize it in 3 dimensions, and navigate around it. You can also fly through your displayed data in 3-D perspective and make movies of your flight.

You can convert 2D points, lines, and polygons to 3D and the 3D work can be saved as an _.sxd_ extension.

Z Values

To work with the 3-D Analyst you need a 3D data set that consists of (x, y, z). The x, y is the location and z represents a value in the attribute other than the location. The z could be elevation of points above sea level, chemical concentration of groundwater, precipitation, or any phenomenon that varies across a specific location. Table 17.1 shows that the x, y is the easting and northing, and the z could represent either the elevation, well depth, or salinity.

Raster

A raster is a way to represent digital images which has a wide range of formats, including the .gif, .jpg, .tiff, .png, .bmp and others. A Raster consists of a matrix of cells (grid) organized into rows and columns. The cell can represent a square kilometer (km^2), a square meter (m^2), or a square cm (cm^2). The square cm represents a more detailed raster with a higher resolution than the bigger cells. The higher the resolution of the raster, the larger the file space is that's taken up by the image.

When a raster image is created, the image on the screen is converted into pixels. Each pixel is assigned a specific value which determines its color. The raster image system uses the red, green, blue (RGB) color system. An RGB value of 0, 0, and 0 would be black, and the values go all the way through to 256 for each color, allowing the expression of a wide range of color values.

Electronic Supplementary Material: The online version of this chapter (https://doi.org/10.1007/978-3-319-61158-7_17) contains supplementary material, which is available to authorized users.

Table 17.1 Groundwater wells location and other information

Well_ID	Northing	Easting	Elevation	Depth	Salinity
840	3872.70	1027.00	626.99	17	456
841	3872.50	1024.20	663.03	38	567
842	3871.50	1021.80	662.91	58	435
843	3928.50	1358.20	662.73	14	289
846	3572.20	941.30	662.38	35	987
847	3554.10	2416.40	664.35	14	888
848	3038.90	964.70	663.61	16	846
853	2501.00	808.80	668.19	55	484
856	2133.00	2387.70	673.90	35	503
857	2129.10	2390.20	673.95	61	359

The x, y coordinates of the image are not stored in each cell, but they calculated from the x, y location of the upper-left cell in the raster. The cell stores only the z value that represents a category of phenomena or quantity such as elevation or salinity, or depth (Table 17.1).

Raster data can be divided into two categories

Image: represents the reflection of light or some other band in the electromagnetic spectrum and can be measured by camera or satellite.

Thematic: represents a category of phenomena or quantity such as elevation or salinity, and has to be measured by a human using an instrument such as a GPS (in terms of elevation). Several samples will be taken and then a surface model will be interpolated using a certain methodology such as statistics or mathematics.

Triangulated Irregular Network (TIN)

The TIN represents a surface as a set of irregularly located points, joined by a line which forms triangles of different sizes. Each triangle point (node) stores the x, y, and z values. The values in a TIN are interpolated from the collected sample points like a raster. The value on a TIN surface will be estimated using the x, y, and z values of each triangle. In addition, the slope and aspect of each triangle face will be calculated. The TIN takes more space, and therefore it is recommended to use them for small areas that require accurate modeling presentation.

3-D Features

The 3-D vector format is like the 2-D vector format, as it can represent any point, line, and polygon. The 3-D vector stores the z value along with x, y coordinate. Each point has only one z value and the line and polygon has many z values, one value for each vertex.

It is recommended adding the ArcScene icon into the desktop.

Lesson 1: Working with 3-D in ArcCatalog

Preview Raster (DEM) and Vector in ArcCatalog

1. Click Start menu/Point to All Programs/ArcGIS/click ArcCatalog
2. Customize menu/Extensions/check the box next to 3D Analyst
3. Close
4. Customize menu/Toolbars/check the box next to 3D View Tools

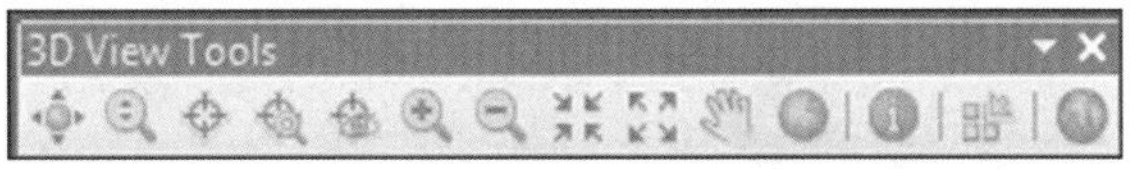

Result: The 3D Analyst extension and the 3D View Toolbar are loaded.

5. Click "Connect To Folder"  icon navigate to the \\Ch17\ folder
6. Click the plus sign next to the **Q1** to open it
7. Q1 contains the DEM of "**Duluth.tif**" and shapefile "**Contour.shp**"
8. Click "**Duluth.tif** " in the Catalog tree and then click the Preview tab
9. The "**Duluth.tif**" is displayed in 2D

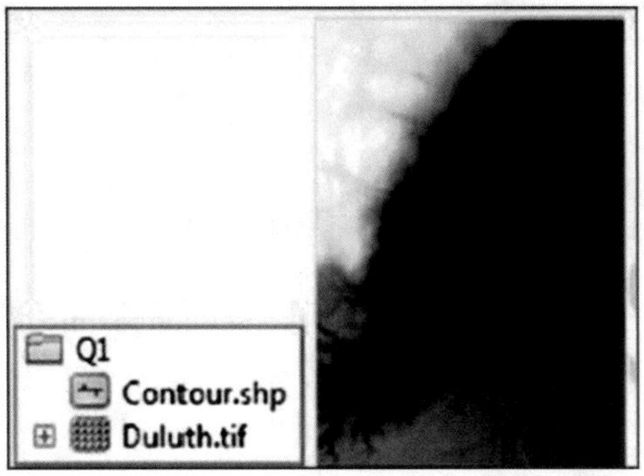

Note: 2D is called orthographic or planimetric view.

10. Click the arrow next to **Preview** at the bottom of the page/select 3D View
11. The DEM "**Duluth.tif**" is displayed in 3D (perspective view)
12. Click the Navigate tool

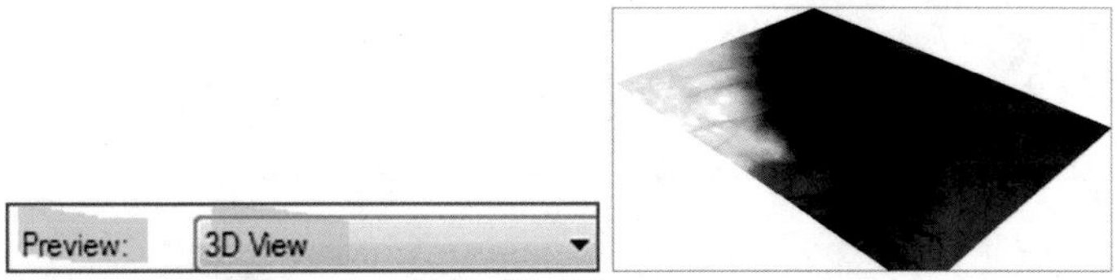

13. Place the cursor over the "**Duluth.tif**" 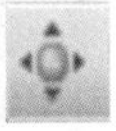
14. Hold down the left mouse button and move it in any direction
15. Click Full Extent 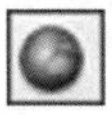 icon on the 3-D toolbar to go back to the original position
16. To Pan, hold both left and right mouse button and drag the cursor in any direction
17. To Zoom in and out hold down the right mouse button and drag it down and up
18. Click "**Contour.shp**" in the Catalog tree, then click the preview tab
19. The "**Contour.shp**" is displayed in 3D View (Preview tab is already set to 3D)
20. Rotate, Zoom In, and Pan the "**Contour.shp**" as in the previous steps.

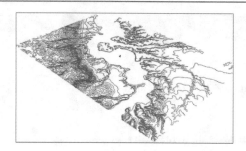

Does the Shapefile Contain 3D Features?

In order to know if the shapefile contain 3D feature do the following

21. Click "**Contour.shp**" in the Catalog tree again
22. Click in the Preview tab above the display map
23. Set the Preview menu at the bottom of the page to "Table"
24. You will notice that the Shape* field does not contain "**Polyline ZM**"

   Note: The "**Polyline ZM**" indicates that the "**Contour.shp**" is a 3D line feature.
   Result: The "**Contour.shp**" is not a 3D feature.

Create a Layer File for the DEM of Duluth in ArcCatalog

This exercise is to create a layer file (.lyr) in order to see the Duluth DEM in 3D. The layer file is not a raster data but a copy of the display instruction.

1. Click "**Duluth.tif**" in the Catalog tree and then click the preview tab
2. Make sure that the Preview menu is set to 3D View
3. R-click "**Duluth.tif**" DEM and point to Create Layer
4. Save it as "**Duluth.lyr**" in the \\Ch17\Result folder
5. Highlight the "**Duluth.lyr**" in the Catalog tree and select Preview tab/select 3D View

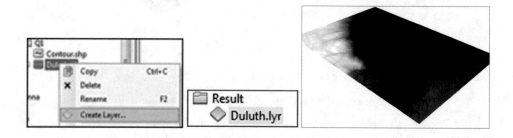

Display Duluth.lyr in 3D Using the Base Height of Duluth DEM

The next step is to display the "**Duluth.lyr**" in 3D to see the overall elevation of Duluth.

Base Height for Duluth DEM
6. R-click "**Duluth.lyr**"/Properties/click Base Height tab
7. Choose *Floating on a Custom Surface*

8. This step will use the elevation values from the Duluth DEM \\Q1\Duluth.tif
9. Below change the custom to 5 (will apply exaggeration to the elevation)/OK

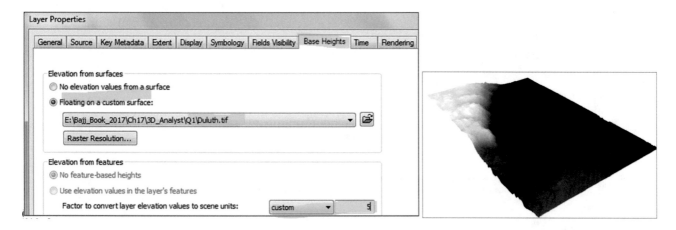

Result: The Duluth file layer is displayed in 3D with exaggeration.

Change the Color of the Duluth Layer File

11. R-click "**Duluth.lyr**"/Properties/Symbology tab
12. Under Shows: make sure Stretched is selected
13. R-click in the Color Ramp window and Uncheck the Graphic View
14. Click the drop-down list and scroll down to select Elevation # 1/OK

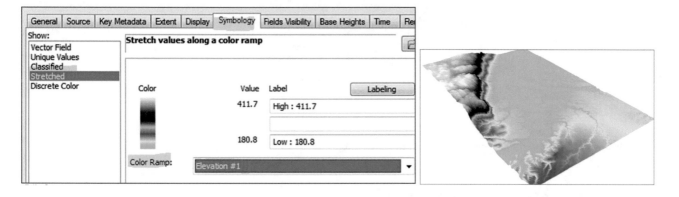

Result: A color added to the Duluth layer.

Add Shading to the Duluth Layer File

16. R-click "**Duluth.lyr**"/Properties/click Rendering tab

Note: The Rendering tab consists of **Visibility**, **Effects**, and **Optimize** frames.

17. Under **Visibility** frame, make sure "Render layer at all times" is checked
18. Under **Effects** frame check "Shade areal features relative to the scene's light position"
19. Under **Optimize** frame, make sure "Cache layer for fastest possible rendering speed"

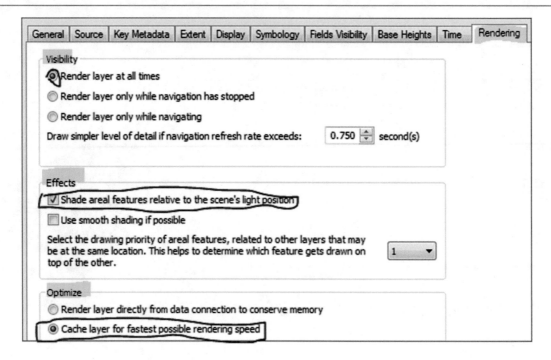

20. Click OK

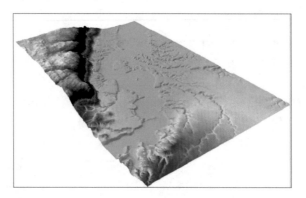

Result: The elevation of the Duluth layer is better presented now.

Create a 3D Thumbnail to the Duluth Layer File

21. In the Catalog tree, click the "**Duluth.lyr**"
22. Select the Preview tab above the display map
23. Click the Create Thumbnail button on the 3D toolbar

24. ▦ Make sure the Preview tab at the bottom is set to 3D
25. Click the Content tab
26. The Duluth Layer File thumbnail is added

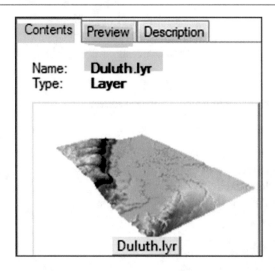

Lesson 2: Working with 3-D Toolbar in ArcMap

You can access 3D Analyst toolbar in ArcMap, but make sure that the 3D Analyst extension is enabled. The Toolbar contains tools you can use on 3D surfaces to perform different functions such as create a contour line, interpolate heights for digitized points, lines, and polygons and much more. These tools work with TIN, terrain dataset, LAS dataset surfaces, and DEM.

While using the 3-D Toolbar, you will perform the following:

- Create Single Contour
- Interpolate Points
- Interpolate Line

1. Launch ArcMap and rename the Layer data frame "**Duluth Height**" Customize menu/Toolbars/3D Analyst
2. Click Add Data and browse to \\Ch17\Data\Q2 folder and integrate City.shp, Duluth.tif, RainStation.shp, and Stream.shp layers
3. Change the symbols, size, and color of the City.shp, RainStation.shp, and Stream.shp layers based on your taste
4. Change the Symbology of the Duluth.tif and apply the Elevation1

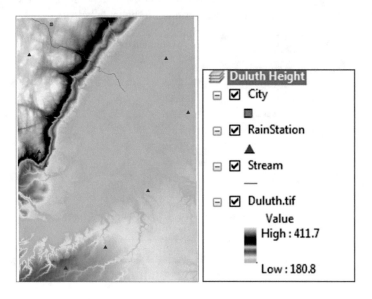

Result: The layers will be displayed and they are symbolized.

Create Single Contour

Contour lines are representing surface elevations on maps. A contour is a line that connect all points with equal height (or other values such as population, chemical parameter).

You can create contour lines for a raster, TIN, or DEM.

In this exercise, you are going to create contour line for the city layer

1. Click the Create Contours ![icon] button on the 3-D Analyst Toolbar.
2. Click the surface at the City layer to create a layer
3. The contour is added as a 3D polyline graphic.
4. The height of the City layer is 380.587 meters (It is written to the status bar).

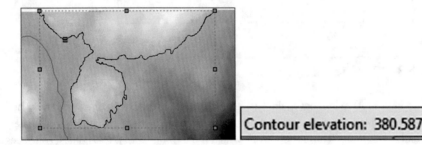

Contour elevation: 380.587

Find the Elevation of the Rain Stations Using the "Interpolate Point"

The tool create a 3-D point by interpolating the elevation from the DEM of Duluth image. In this exercise you want to find the elevation of the RainStation layer using the DEM of Duluth.

5. R-click RainStation layer in the TOC/Label Features
6. Click the Interpolate Point ![icon] button on the 3-D Analyst Toolbar.
7. Click the symbol of the RainStation that has a label 1
8. A point appears above the symbol and the height of the Rainstation is 397.234 meters written to the status bar.
9. Continue identifying the elevations of the RainStation layer and fill the table below

RainStation No	Elevation (m)
1	397.234
2	
3	
4	
5	
6	
7	
8	
9	

Point Z: 397.234

Create a Profile Graph for the Stream Using the "Interpolate Line"

The Interpolate tool allows you to create a 3-D graphics on a DEM using the "Interpolate Line" on the 3-D Analyst

10. Insert Data Frame and call it "Vertical Profile"
11. Drag the Duluth.tif and Stream layer from the Duluth Height data frame into the "Vertical Profile" data frame
12. Make the width of the Stream layer 2

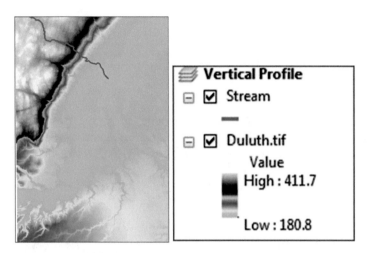

13. Click the Interpolate Line [icon] button
14. Then click the upper end of the Stream layer and digitize a line along the Stream flow
15. When you are finished adding vertices to the line, double-click to stop digitizing.

16. Then click the "Create Profile Graph" [icon] button.

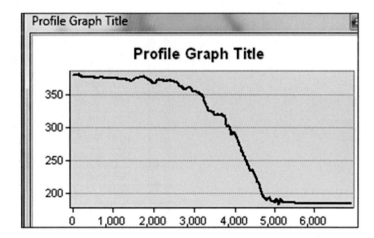

Result: The vertical profile display.

Lesson 3: Working with ArcScene

ArcScene is a 3D visualization application that allows you to view your GIS data in three dimensions. It allows the user to integrate many layers of data into a 3D environment. Data can be displayed in 3D by using height information from raster elevation or a field in the attribute table. Data with different spatial references will be projected to a common projection, or data can be displayed using relative coordinates only. ArcScene also has the capability to utilize many analytical tools and functions.

Scenario 1: You have been asked by your advisor to integrate various types of data and create a TIN in order to provide a presentation about the water resources in Dhuleil area in 3-D setting.

Create TIN from Contour Line

A triangulated irregular network (TIN) is a digital data structure used in GIS for the representation of a surface. The TIN is a vector-based digital data and is constructed from features, such as points, lines, and polygons that contain elevation information. There are different methods of interpolation to form these triangles, such as Delaunay triangulation or distance ordering. ArcGIS supports the Delaunay triangulation method.

ArcScene can be launched from the Start menu and it reside in ArcGIS and you can start the program also from the

ArcScene  button on the 3D Analyst toolbar in ArcMap

1. Click Start menu
2. Point to program then ArcGIS
3. Click on ArcScene to launch it
4. Click Add Data button, and browse to \\Data\Q3 folder and integrate **Contour.shp**
5. R-click the **Contour** layer and Open Attribute Table

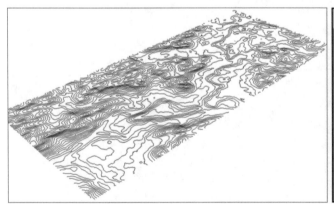

Contour			
FID	**Shape**	**ID**	**Elevation**
123	Polyline	112	450
115	Polyline	103	460
135	Polyline	118	460
144	Polyline	123	460
106	Polyline	992	470
149	Polyline	131	470
105	Polyline	974	480
153	Polyline	141	480

Note: The table has an "Elevation" field that will be used to create the TIN.

6. Close the Contour attribute table Click on the ArcToolbox button on the Standard Toolbar to launch it

7. Click ArcToolbox/3D Analyst tools/Data Management/TIN
8. D-click Create TIN
9. Fill the Create TIN dialog box as follow:
10. Output TIN: \\Ch17\Result\Dhuleil_TIN
11. Coordinate System browse to \\Data\Q3 and import from Contour.shp/OK
12. Input Feature Class: Contour
13. Make sure the Height Field is **Elevation** and SF Type is **Hard_Line**
14. Click OK

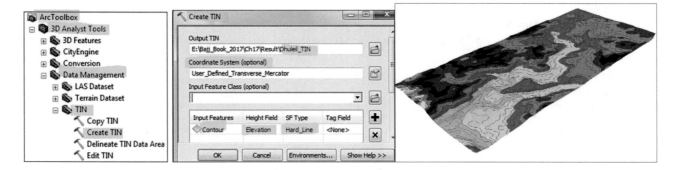

Result: **Dhuleil_TIN** is created and added to TOC.

Change the Symbols of the TIN

15. In the TOC/R-click "Dhuleil_TIN"/Properties/Symbology/
16. Under Show: uncheck Edge types and Elevation and click Add button
17. Highlight "**Faces with the same symbol**"/click Add then Dismiss
18. Click the Symbol color and click Dawn Ivory/OK/OK

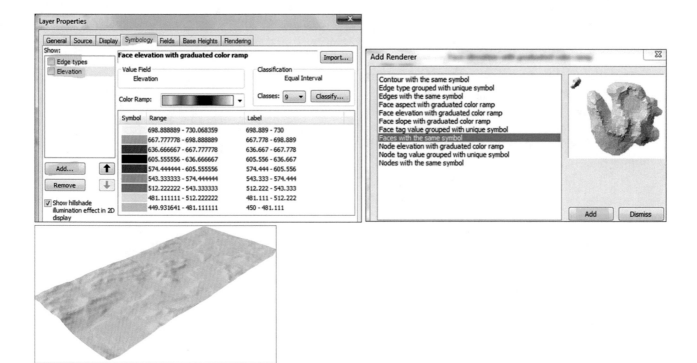

Result: **Dhuleil_TIN** has now one color.

Drape and Extrude Layers onto Dhuleil_TIN

In this exercise, you will drape the different layers over the TIN. The layers don't have an elevation field in their attribute tables, so you will take the base height from Dhuleil_TIN and apply them to the layers.

19. Click Add Data button and browse to \\Data\Q3 and add **Farm**, **ObserbationWell**, **Street**, **SupplyWell**, **Valley**, and **WWTP** layers
20. In TOC/R-click **SupplyWell**/Properties/click Base Heights tab
21. Under "Elevation from Surfaces" check "Floating on a custom surface"
22. Choose Dhuleil_TIN
23. Click the Extrusion tab/check "Extrude features in layer"
24. Click the Calculator and enter the following expression "**[Height] *10**"
25. Click Ok/Apply/OK
26. Change the symbol of SupplyWell into "Dot 4" and select Cretan Blue color/click OK
27. Repeat the previous steps for the ObservationWell
28. Change the symbol of ObservationWell and into "Dot 4" and select Apatite Blue color

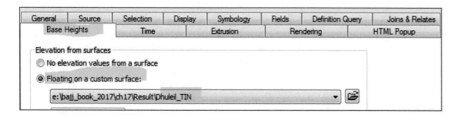

29. In TOC/R-click **Valley**/Properties/click Base Heights tab
30. Under "Elevation from Surfaces" check "Floating on a custom surface"
31. Choose Dhuleil_TIN
32. Change the **Valley** to the River symbol and make the width 2
33. Repeat the previous step for the **Farm**, **Street** and **WWTP** and change the color into Light Apple, Black and Ultra Blue respectively

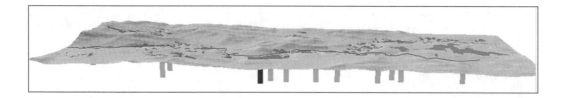

34. Click Add Data button and browse to \\Data\Q3 and add **Building**, **GasStation**, and **Tree** layers.
35. In TOC/R-click **Building**/Properties/click Base Heights tab
36. Under "Elevation from Surfaces" check "Floating on a custom surface"
37. Choose Dhuleil_TIN

38. Click the Extrusion tab/check "Extrude features in layer"
39. Click the Calculator and enter the following expression "**[Height] *20**"
40. Click Apply/OK
41. Repeat the previous step for the GasStation and extrude it by multiply the height by 10.
42. In the TOC/uncheck the Contour layer
43. D-click anywhere in the blank area of a toolbar and click 3D Effects/Close
44. On the 3D Effects toolbar/click the Layer-drop-down and select "**Dhuleil_TIN**"
45. Click the Layer Transparency [button] button to change the it to 30%
46. Zoom to the building and GasStation layers

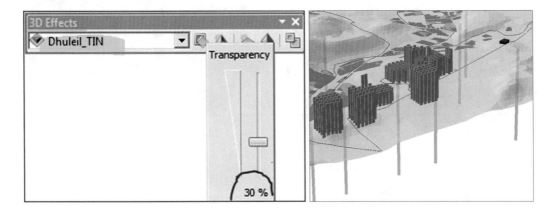

Result: The 3D Effects toolbar will be displayed, and the Layer Transparency will be changed to 30%.

Applying the 3D Symbol to the Tree Layer

Points in ArcScene can be symbolized as 3D markers and you can choose the 3D markers from the existing styles in ArcScene.

47. In TOC/R-click **Tree**/Properties/click Base Heights tab
48. Under "Elevation from Surfaces" check "Floating on a custom surface:
49. Choose Dhuleil_TIN
50. Click Symbology tab/click the Symbol button/click Style References button/select 3D Trees/click OK/select the Pine symbol/make the size 25/click OK/OK

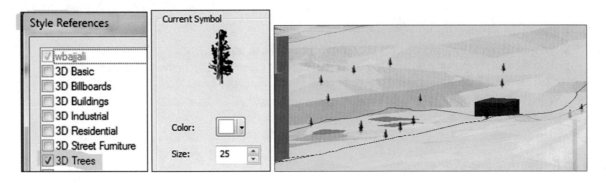

Vertical Exaggeration

ArcScene allows the user to exaggerate the vertical appearance and also change the elevation of the surface by any number. The exaggeration can provide a better visual effect and smooth the look of the surface, but does not affect the analysis.

51. R-click Scene Layers/Scene Properties dialog box display
52. Change the Vertical Exaggeration onto 2
53. Apply/OK

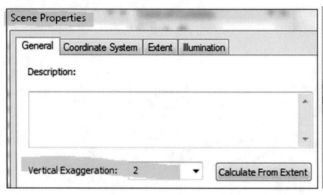
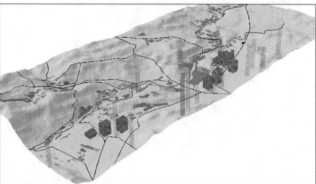

Navigate and Fly

54. Click Full Extent 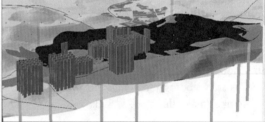 button on the Tools toolbar

55. On the Tools toolbar/click the Navigate button (1st icon)

56. Click on Dhuleil_TIN and drag the map to view the scene from different positions

57. Click the Zoom In button and zoom in to a small area around the Building layer

58. Click the Full Extent button

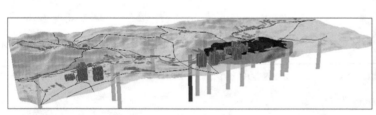

59. From the Tools toolbar/click the Fly button
60. Click in the middle of the map with the Fly button and click again to start a flight
61. Slowly move the mouse up, down, to left, and to right in order to move the map to these directions
62. Click the left mouse to increase the speed, the right mouse to decrease the speed
63. Press ESC key on the keyboard to stop the flight
64. Zoom to Full Extent

Create an Animation and Video

65. D-click in the blank area of a toolbar
66. Check the Animation and click Close button

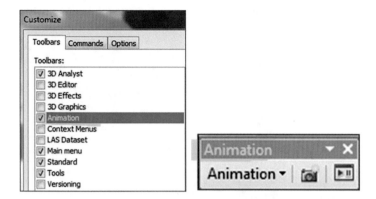

67. On the Animation toolbar/click the Open Animation Control  button

Result: The Animation Controls toolbar display.

68. Zoon In in the Middle of the Scene
69. Click the Record button
70. Click the Play button
71. Click the Fly button and create a fly through the whole study area and when finish click the ESC key to stop the flight
72. On the Animation toolbar/click Animation/Save Animation File
73. Browse to \\Result folder and save the animation as **Dhuleil.asa**/click OK
74. On the Animation toolbar/click Animation/Load Animation File/
75. Browse to \\Result folder, highlight **Dhuleil.asa** and click Open
76. From the Animation Controls toolbar/click Play button

77. Zoom to Full Extent
78. On the Animation toolbar/click Animation/Export Animation
79. Browse to \\Result folder and save the animation as **Dhuleil.avi**/click Export
80. The Video Compression dialog box display
    a. Compressor: Microsoft Video1
    b. Compression Quality: 100
81. Click OK
82. Launch PowerPoint
83. Insert menu/Video/Video on my PC browse to \\Result
84. Highlight Dhuleil.avi/Insert
85. In PowerPoint/Click play to watch the video

Time Tracking

Time tracking is a visual representation that uses the time field to show how the events are changing over time. The Plume attribute table layer contains a time field, which allows user to visualize the events at various locations over time.

86. Activate ArcScene and Zoom to Full Extent
87. Add Data and browse to \\Data\Q4 folder and add **Plume** layers.
88. In TOC/R-click Plume/Properties/click Base Heights tab
89. Under "Elevation from Surfaces" check "Floating on a custom surface"
90. Choose Dhuleil_TIN and click OK
91. Change the Color of Plume to "Mars Red"
92. OK
93. R-click Plume/Properties/Time tab
    a. Check "Enable time on this layer"
    b. Layer Time: Each feature has a single time field
    c. Time Field: Coll_Date
    d. Click Calculate
    e. Time Step Interval: change it into "1 Year"
    f. Check Display data cumulatively
94. OK

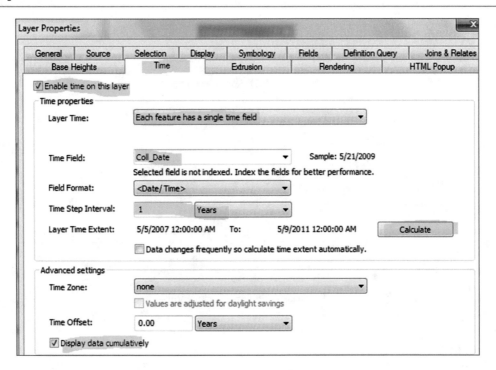

Note: Once you clicked OK the Time Slider  on the Tools toolbar will be activated.

95. Click on the Time Slider [icon] button on the Tools toolbar

96. The Time Slider display

97. Click on Disable time on map [icon] button (1st icon on Time Slider

98. Click on the Options [icon] button (2nd icon)
99. Click Time Display tab/Change it as follow
100. Time Step Interval: 1 Year
101. Time window: 1.0 Year
102. Click Playback tab
103. Check "Display data for each timestamp"
104. Speed: move the speed selector to the middle between slower and faster
105. After playing once: Repeat
106. Check Refresh the display when dragging the time slider interactively
107. OK

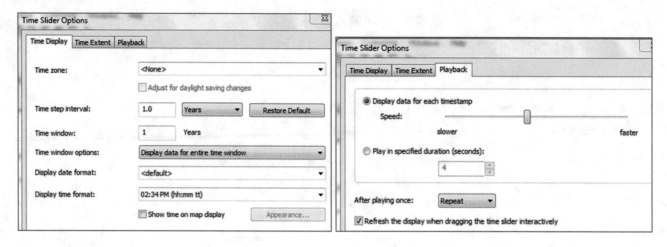

108. Click Play icon

109. Click the Export to Video 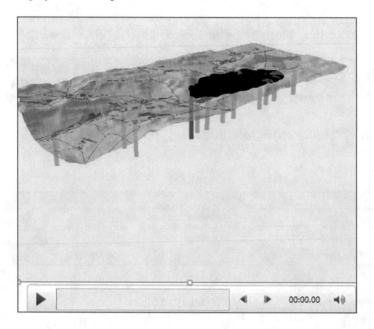 button (3rd icon) on the Time Slider toolbar

110. Save it in \\Result folder as **Plume.avi** then click Export

111. Accept the default and click OK

112. Launch PowerPoint

113. Insert the **Plume.avi** and play it as in the previous section

Result: The video will play.

Introduction

Mobile GIS is simply a field GIS, where the user use a mobile devices. The device is armed with a software called ArcPad, which that can perform special jobs in the field. The ArcPad mobile GIS can be used away from the office to capture data in GIS format and when user return to the office, the data can be integrated into ArcGIS as a complete database. The mobile ArcGIS has been used by a wide range of professional users of different disciplines, including; engineering, geology, farming, government, environment, and others. The mobile ArcGIS provides many benefits because of the following:

1. Users can capture and obtain data in real time.
2. Inspectors can update records in the field efficiently and quickly.
3. Incidents can be documented and reported from the field to the office using wireless communications.
4. Provide direct data to the decision-making.

The most common functions that Mobile GIS have been designed to serve are:

1. Field data capture, either by real-time editing of supplied digital maps, or by surveying and mapping in the field using GPS input to create new features (e.g. a manhole checking, or capturing a tree location)
2. Navigation with the GPS, where you can find your current location on a moving map. Display and use GoTo functions to locate destination points (e.g., relocating a well location)
3. Check to validate of an existing feature or updating (e.g., checking if geological fault exist)
4. Geocoding land, property, and infrastructure, then integrating this information with point-of-interest databases (e.g., a utility inspector conducting an asset audit)
5. Integration with enterprise databases from the field (e.g., a data librarian updating the location of Wi-Fi hot spots in a database)

Scenario 1: You are going to use Trimble Juno T41GPS as a receiver to capture field data. Your task will be achieved in two parts. Part one you will perform in the office using your ArcGIS to create a project containing the geodatabase and image need for field collection. Part 2 is to capture the locations of trees, sidewalks, and buildings on the main UWS Campus.

Part 1: Prepare the Data in the Office for ArcPad to Perform Field Work

The work consist of the following stages that needed to prepare the data for the field work

1. Convert the image to make it usable in Trimble Juno T41GPS with ArcPad.
2. Prepare the data in ArcGIS and integrate it in Mobile GIS by doing the following.
   a. Create a file geodatabase in ArcGIS call it **Campus**, to contain existing data for the field area.

Electronic Supplementary Material: The online version of this chapter (https://doi.org/10.1007/978-3-319-61158-7_18) contains supplementary material, which is available to authorized users.

b. Create a feature dataset in the Campus geodatabase containing empty polygon, line, and point feature classes for field data entry.

c. Create a project containing all data needed for field data collection.

Prepare Raster Data for ArcPad Desktop

Convert Raster Format into GEOTif to Make It Usable in ArcPad Desktop

ArcPad supports raster image display including numerous industry-standard formats. In this exercise you are going to work with TIFF format (*.tif). We are going to use an aerial photograph downloaded from city of Superior registered in Douglas County projection (dc4). The image will be clipped to include only the UWS campus. Then it will be projected using the Projection Raster tool to project the image into latitude-longitude using the world datum (WGS84) and we are going to call it **UWS_GCS.tif**. After projecting the image we are going to do the following steps in order to make it usable in **ArcPad** desktop only. This step is important to perform in order to work with Trimble in the field.

1. Start ArcMap
2. Browse to \\Ch18\ Data\image folder and integrate **UWS_GCS.tif**

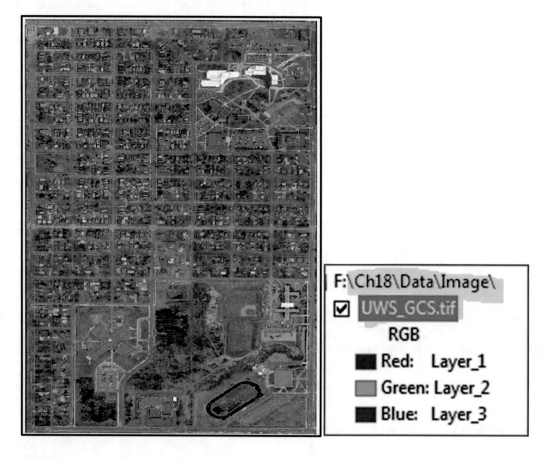

Note: This image has a Tiff format and three bands.

3. Click File menu/Export Map
4. Save in: \\Ch18\ Data\ArcPad folder as Campus.tif
5. File name: Campus.tif
6. Save as type: TIFF (*.tif)
7. Click General tab/check "Write World File"
8. Click Format tab/check "Write Geo TIFF Tags"

9. Click Save

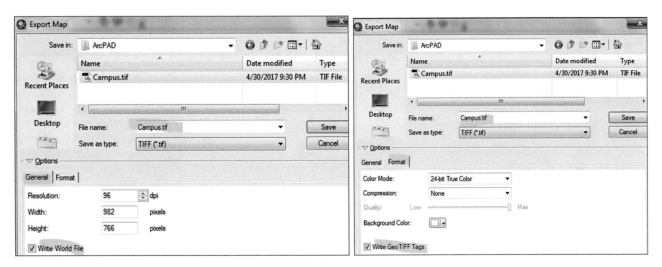

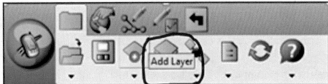

10. Start ArcPad on your computer desktop

11. Click New map, ArcPad will open\ Make sure the Main Tools  is checked

12. Click Add Layer button

13. Browse to \\Data\ArcPAD folder

14. Check the Campus.tiff

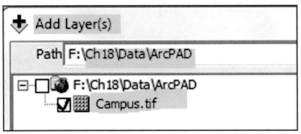

15. OK

Result: The **Campus.tiff** is displayed in ArcPad, this means this image is now ready to be integrated into a mobile GIS.

Prepare File Geodatabase in ArcGIS and Integrate It into Trimble Juno T41GPS

The **UWS_GCS** image for this project is available in tiff format, which is an acceptable format to work with ArcPad. In this exercise you are going to build a file geodatabase and a feature dataset and register it in latitude-longitude using the WGS84 datum. After creating the Feature Dataset, you are going to create an empty polygon, line, and point feature classes. Then you will integrate the file geodatabase and the image UWS_GCS.tiff into the mobile GIS (Trimble Juno T41GPS). After this step, you will use the ArcPad software in Trimble Juno T41GPS to collect the data in the field using the **UWS_GCS.tiff** image as a background reference.

Create File Geodatabase in ArcMap

1. In ArcMap/open the Catalog window browse to Ch18
2. R-click Ch18\New\File geodatabase\ call it **UWS**
3. R-click **UWS** File geodatabase\New\Feature Dataset\call it **Campus**
4. Click Next
5. Select Geographic Coordinate Systems/World/**WGS 1984** coordinate for the **Campus Feature Dataset**
6. Next/Next/Finish

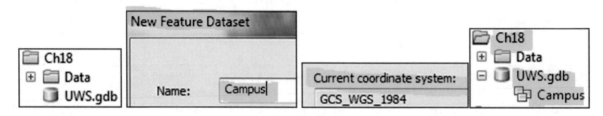

Result: The Campus Feature Dataset is now created and it is empty.

Next Step: Create polygon, line and point feature classes inside the **Campus Feature Dataset**.

Create a Polygon Feature Class
7. R-click **Campus Feature Dataset**/New/Feature Class
8. Name: **Building**
9. Type of feature stored in this feature class: **Polygon Features**

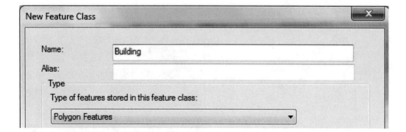

10. Next/Next/
11. Under Field Name/type NAME/
12. Under Data Type/select Text
13. Under Field Properties/Length = 16
14. Finish

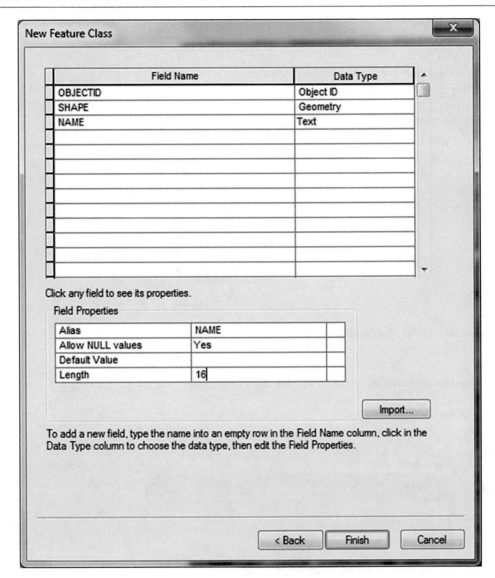

Result: The **Building** feature class is added to the TOC, and the attribute table is empty.

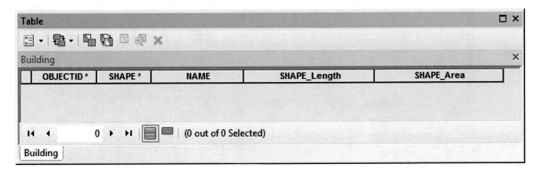

Create a Sidewalk Feature Class

15. R-click **Campus Feature Dataset**/New/Feature Class
16. Name: **Sidewalk**
17. Type of feature stored in this feature class: **Line Features**
18. Next/Next/Finish

Result: The **Sidewalk** feature class is added to the TOC, and the attribute table is empty.

Create a Tree Feature Class

19. R-click **Campus Feature Dataset**/New/Feature Class
20. Name: **Tree**
21. Type of feature stored in this feature class: **Point Features**
22. Next/Next/Finish

Result: The **Tree** feature class is added to the TOC, and the attribute table is empty.

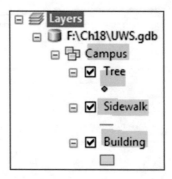

23. Integrate **Holden.tif** into ArcMap from\\Ch18\Data\Image

Import the Image (Holden.tif) into the UWS File Geodatabase

24. R-click UWS.gdb/Import/Raster Datasets
25. Input Rasters: \\Data\Image\Holden.tif /Add
26. OK

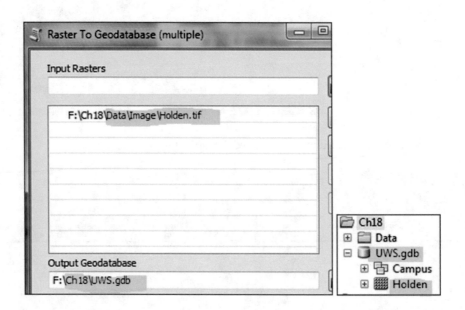

Result: The image is now part of the UWS file geodatabase.

27. In ArcMap/File/New/Click Blank Map/OK
28. Click No (Save changes to Untitled?)
29. Integrate the **Holden.tif** from UWS.gdb, and **Building**, **Sidewalk**, and **Tree** feature classes from Campus Feature Dataset
30. In ArcMap Save the map as **UWS.mxd** in \\Data\ArcPad folder (you are going to use it later on).

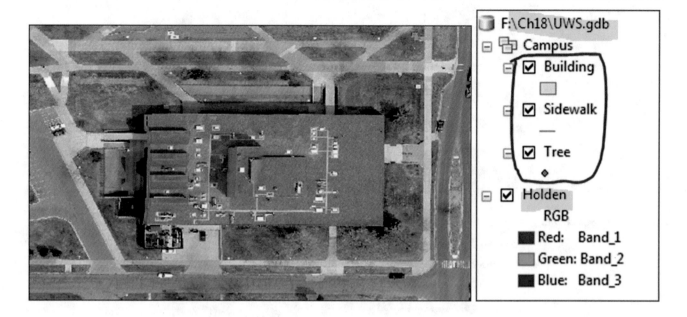

Using ArcPad Data Manager Tool to Transfer Data from ArcGIS into the Mobile GIS

At this stage, you are going to use the **ArcPad Data Manager** tool to transfer data from the **Campus Feature Dataset** and the **Holden.tif** image from **UWS.gdb** in ArcGIS into a mobile device (Trimble Juno GPS) for use with ArcPad in the field. All the data that is copied out, is automatically converted into a format that ArcPad can use.

31. In ArcMap/Customize/Extensions/make sure that ArcPad Data Manager is checked.

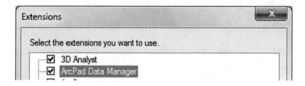

32. Customize/Toolbars/check ArcPad Data Manager

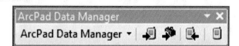

Result: The **ArcPad Data Manager** tool is displayed.

   Before you start transferring the data make sure that all the data in the Campus Feature Dataset is displayed in the TOC (**Holden.tif**, **Building**, **Sidewalk**, and **Tree** feature classes).

33. Click Get Data For ArcPad button (first icon)

34. Next
35. Click the **Action** menu header

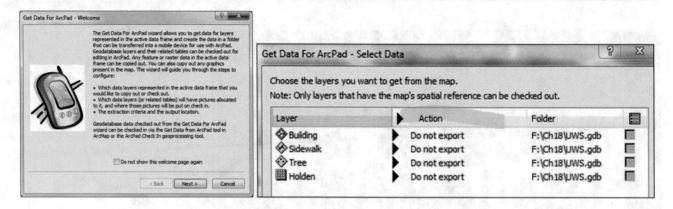

Note: In the "Select Data" dialog box, there are several options for how to export data under the "**Action**" column.

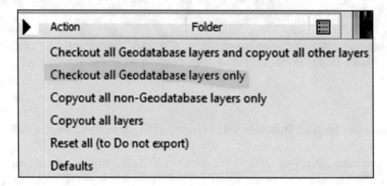

The menu has two options, the first is **Checkout** and the second is **Copyout**. A "**Checkout**" is only allowed for Geodatabase layers. **Copyout** is only for shapefiles or other layers not in a geodatabase.

36. Click Checkout all Geodatabase layer only

Result: Under **Action** the <u>Tree</u>, <u>Sidewalk</u>, and <u>Building</u> will displayed as **Checkout Data**.

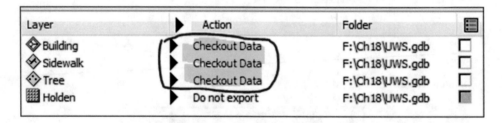

37. Click the **Holden** image under **Action**, and select **Export as Background TIFF**

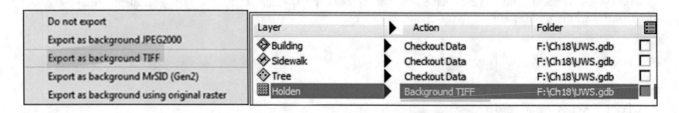

38. Click Next

39. Click Next (because there are no pictures associated with the files)

40. Specify a name for the folder that will be created to store the data: **HoldenForArcPad**
41. Where do you want this folder to be stored: \\Ch18\Data\ArcPad folder
42. Map Name: **Holden**
43. Next
44. Make sure to check "Create the ArcPad data on this computer now"/Finish/OK

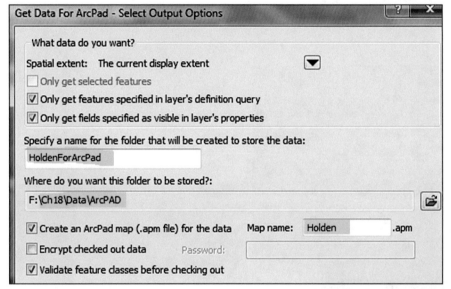

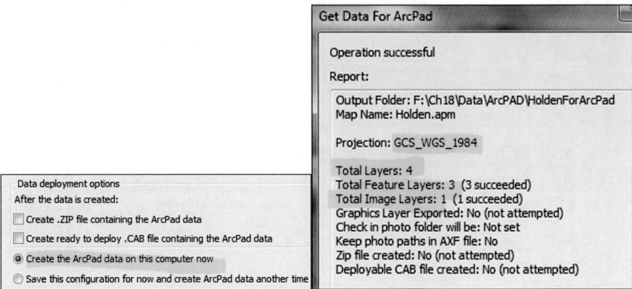

Result: A message will display stating that 4 total layers succeeded (3 layers and 1-image) and they are registered in latitude longitude and associated with WGS84.

If you open the folder to \\Ch18\Data\ArcPad folder you will see **HoldenForArcPad**. This folder contains the files that are needed to be transformed into the Juno Trimble.

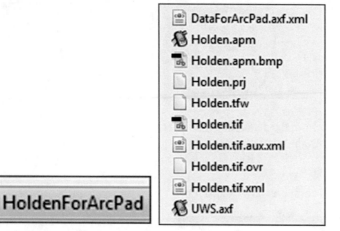

45. Connect to Trimble Juno

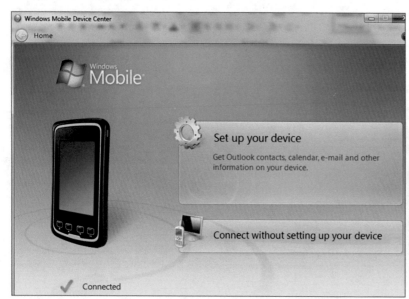

46. Click File Management/click "Browse the content of your device"

Result: You see two folders; the main memory and storage memory.

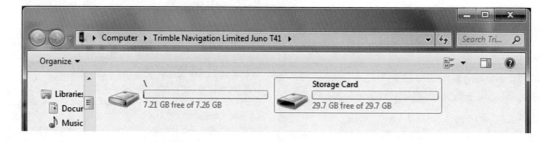

47. Open the Storage Card and copy the **HoldenForArcPad** from your PC to the Storage Card

Result: The **HoldenForArcPad** is copied in the Storage Card.

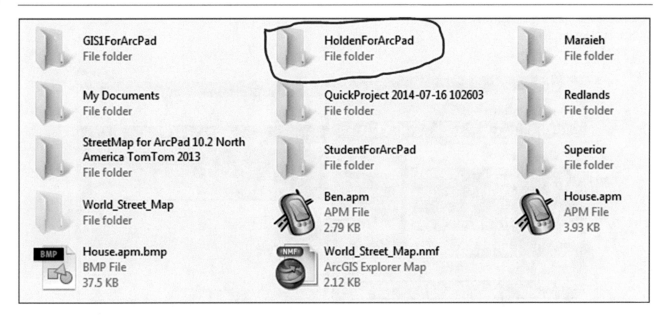

Starting ArcPad in Trimble Juno T41

48. Switch on Trimble Juno
49. On the Window Mobile, start ArcPad (Start/Program/click ArcPad)

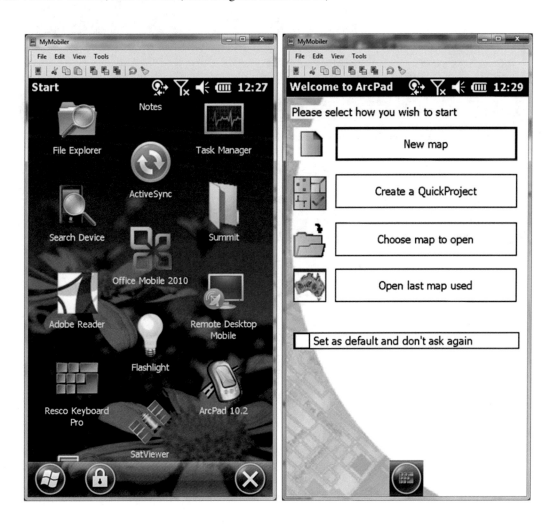

50. Choose the option 'Choose Map to Open'
51. Select "Holden.apm"
52. Click OK

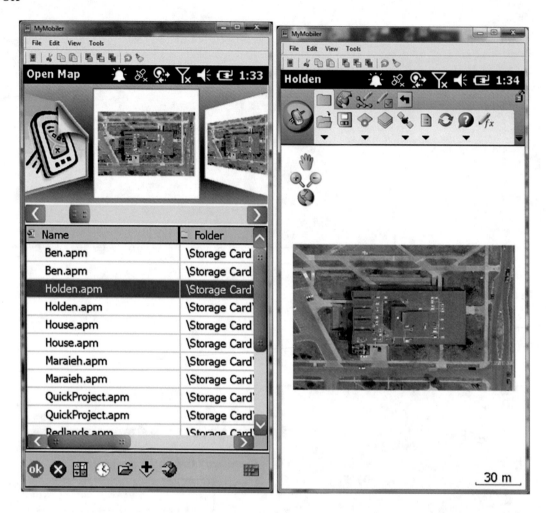

Result: Now the map is displayed and will be used as a reference to digitize the Building, Sidewalk, and Tree layers.

Overview of ArcPad toolbars:
ArcPad tools are organized onto five toolbars, and the user can switch between them. The most important tools are the first four tools, as they are relative to this exercise. Here is a brief description for the four tool bars.

1. *Main* toolbar

2. *Browse* toolbar

3. *Drawing* toolbar

4. *Quick Capture* toolbar

5. *Navigation* toolbar

I. The Main toolbar consists of nine buttons and they are

1. **Open map** ☝️ (Opens an ArcPad map, a file with a *.apm extension)

2. **Save map** 🔲 (Saves the current ArcPad map)

3. **Add layer** ⊕ (Adds one or more layers to the current map)

4. **Table of contents** ▥ (Opens the Table of Contents dialog box)

5. **GPS Active** ⌁ (Activates or deactivates the GPS)

6. **Options** ⦙ (Opens the ArcPad Options dialog box)

7. **Refresh** ⟳ (Redraws the map)

8. **Help** ❓ (Provide ArcPad help)

9. **Quick Fields** ƒx (Turn Quick fields On for this map)

II. The Browse toolbar consists of seven buttons and they are

Zoom In ⊕ (Zooms in on the map using the pen)

1. Zoom to Full Extent 🌐 (Zooms to the full extent of the map)

2. Go Back to Previous Extent ◀ (Zooms back to the previous extent that you were viewing)

3. Identify ⓘ (Activates the Identify tool)

4. Find 🔭 (Opens the Find tool)

5. Clear Selected Feature ▨ (Clears currently selected features)

6. Quick Draw ✎ (Toggles the Quick Draw mode. This mode changes the way the map is displayed, uses less memory, and may improve performance of your map)

III. The Edit toolbar consists of eight buttons and they are

1. Start/Stop Editing ✏️ (This tool is used to select the feature that needs to be edited)

2. Select ▧ (Activates the Select tool)

3. Point ● (Activates the point feature for data capture, you can also activate line and polygon)

4. Capture Point using GPS ⚡ (Capture Point using GPS captures a point feature in the editable point layer using the current GPS position)

5. Add GPS Vertex ⚡ (Captures a single vertex in the current line or polygon feature using the current GPS position)

6. Add GPS Vertices Continuously ✕ (Continuously captures vertices in the current line or polygon feature using the current GPS position)

7. Feature Properties ▦ (Opens the Edit Form or Feature Properties dialog box for the selected feature)

8. Offset Point ➘ (Activate offsets for point data capture)

QuickCapture toolbar

The QuickCapture toolbar, is a dynamic toolbar and its purpose is to provide you with a one-click ability to create new features in your map. You have to consider the following:

If you open ArcPad with no data, the **QuickCapture** toolbar displays only the Capture Photo with GPS.

If ArcPad open with data, the **QuickCapture** toolbar displays the Capture Photo, Capture Point, Capture Line, and Capture Polygon.

If the GPS is on and in the case of capturing a line or polygon features, tapping a tool in the QuickCapture creates the first vertices in the feature. To continue or complete the feature, use the tools on the Command bar. The Command bar consists of the following buttons

The Command bar consists of the following buttons:

1. Save Geometry Changes (Saves geometry changes to an existing feature)

2. Proceed to Attribute Capture (Ends the geometry capture of a new feature and proceeds to capture the feature's attributes)

3. Undo (Undo's the last edit made to a feature)

4. Pen Toggle (Activates or deactivates use of the pen or mouse for capturing features)

5. Cancel (Cancels edits to an existing feature's geometry or cancels the capture of a new feature)

6. Add GPS Vertex (Captures a single vertex in the current line or polygon feature using the current GPS position)

7. Add GPS Vertices Continuously (Continuously captures vertices in the current line or polygon feature using the current GPS position)

GPS Setting in Trimble

Before activating the GPS in the Trimble Juno T41, you need to set the GPS communication parameters in ArcPad to match the parameters set on the GPS receiver in Trimble. The most common communication parameters, are located in the GPS page of the GPS Preferences dialog box. You can use the Find GPS tool to search for your GPS, if you do not know which port on your mobile device your GPS is connected to. However, you will need to make sure that your GPS is connected and turned on, in order for the Find GPS tool to detect your GPS. The Port and Baud drop-down list options will be updated as the Find GPS searches for a connected GPS.

Example: In this device, you will see the following setting:

Protocol: NMEA 0183
Port: COM2
Baud: 4800

Note: Refer to your GPS receiver's manual for information on how to set the output GPS protocol and port communication parameters on the GPS receiver.

Activate the GPS

Before we start capturing the data in the field, you will have to activate the GPS and make sure that the GPS is receiving signals from the satellites.

53. Tap the arrow below the GPS Position Window button to display the dropdown list
54. Tap GPS Active

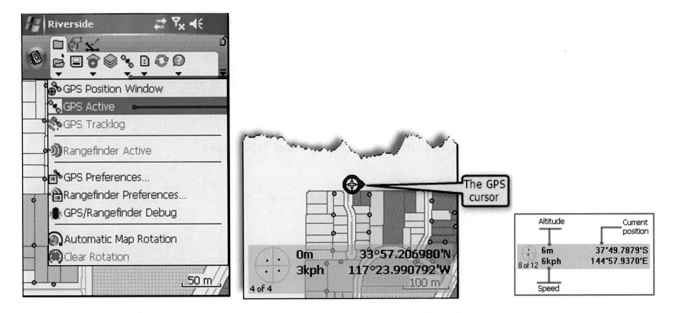

Result: The GPS cursor is displayed as a red circle with a cross inside it, and a small tail. The location (latitude–longitude), the altitude, and speed are also displayed.

Part 2: Capturing Data Using Trimble and ArcPad in the Field

There are two ways to start capturing the data.

I: Capturing Data using the Designed File Geodatabase
II: Quick Project

I: Capturing Data Using the Designed File Geodatabase

This is done if you are in an area and you would like to capture its features using the ArcPad. You should create a file geodatabase and create a different feature class for the feature that you want to capture. For example at the beginning we created three feature classes (**Tree, Sidewalk, and Building**) in the UWS file Geodatabase. Now you are going to capture the three feature classes and use the image (**Holden.tif**) as a background to guide you in the field. The image below shows the area that you are going to capture.

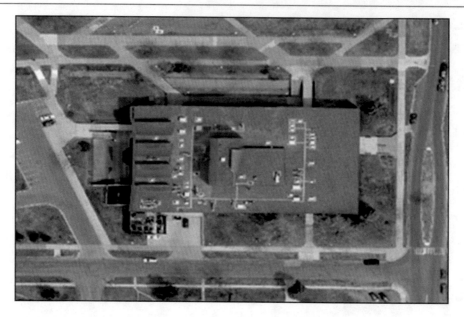

Capturing the Building, Sidewalk, and Trees in the Designated Study Area

Capturing the Holden Building

In order to capture the Holden building, the GPS needs to be activated before any of the GPS buttons on the Edit toolbar are enabled.

1. Click the "Drawing Tools" tab.
2. Click the "Start/Stop Editing" button and select the **Building** that you want to edit
3. Click the "Point Drop-Down List" button, and select **Polygon**.

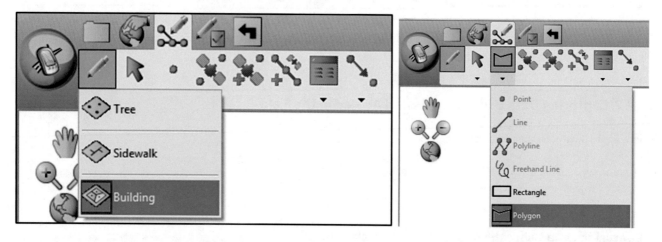

Note: To start capturing the Holden building, you can use either the Add GPS Vertex (Captures a single vertex), or Add GPS Vertices (continuously capture vertices). The Add GPS Vertex and Add GPS Vertices Continuously buttons, are only enabled when the polygon feature type has been selected.

4. Stop at the corner of the building, click Add GPS Vertices and walk around the building. This will capture the perimeter of the **building**
5. Tap the **Proceed tool** on the Command bar when you finish adding your **Building**
6. The Building dialog box opens so that you can immediately enter information about the new Building
7. Type Name: Holden
8. Tap OK

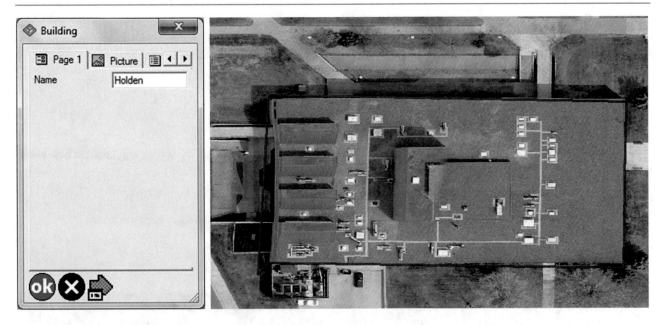

Result: The Holden building is captured and it is shown in red. The boundary of the building is not exactly delineated due to GPS accuracy. This offset will be corrected later on in the office using ArcMap.

Capturing the Sidewalk

The sidewalk around the Holden building needs to be captured using the GPS receiver like in the previous step. With the GPS activated, you will walk the length of the new sidewalk segment, and capture vertices for the length of the new sidewalk.

1. Tap the Start/Stop Editing button.
2. Tap the Sidewalk layer to select it for editing
3. Tap the Polyline feature tool (the Add GPS Vertex & Add GPS Vertices Continuously buttons are enabled)
4. Tap the Add GPS Vertex button, to capture a new vertex. Continue to do so until you have reached the end of the extended sidewalk

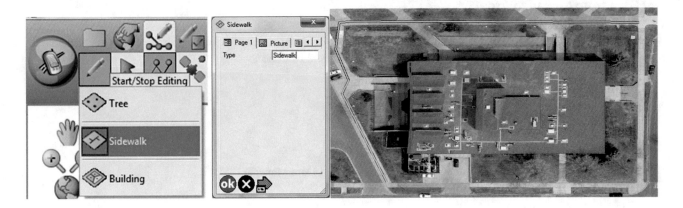

Result: The Sidewalks to the north, and to the west of the Holden building, are captured. It is shown in yellow.

Capturing the Trees

The trees around the Holden building, need to be captured using the GPS receiver like in the previous step. With the GPS activated, you will stand at each tree and use the "Capture Point" button to capture the trees.

1. Tap the Start/Stop Editing button.
2. Tap the Tree layer to select it for editing
3. Tap the Point button on the Edit toolbar

Note: The Point button is now active, and ArcPad is in point-capture mode. Any tap on the screen will create a new point feature at the corresponding coordinates.

4. Tap the map at the location of a tree where you want to create the new point feature.
5. The Feature Properties dialog box is automatically displayed after the new point feature has been created.
6. Type: Tree 1
7. Tap OK
8. Continue capturing the rest of the trees

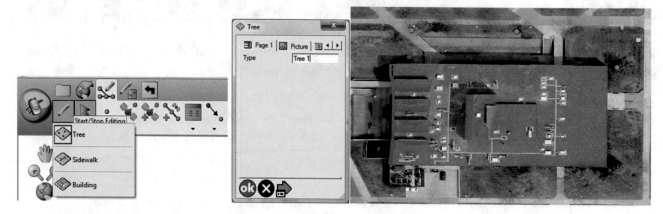

Result: Three Trees in northern part of the Holden building are captured and it's shown in green.

Get Data from ArcPad to ArcGIS

This tool allows you to check-in the edits made in ArcPad, back into the geodatabase from which the data was checked out.
   To get the data from ArcPad to ArcGIS, the following steps should be done:

- The Trimble device should be connected to the PC
- You should copy the HoldenForArcPad from Trimble to your PC

1. Launch ArcMap and open the **UWS.mxd**
2. Click "Get data From ArcPad" button (4th icon) on the ArcPad Data Manager
3. Click **Add ArcPad AXF file**
4. Browse to \\HoldenForArcPad in your PC folder
5. Highlight the **UWS_gdb.axf** and click Open
6. The **UWS_gdb.axf** will be integrated (**Tree, Sidewalk, Building**)
7. Click Select All button (the **Tree, Sidewalk, Building** will be checked)
8. Click the Check in button then
9. Click Yes to check in selected items
10. A dialog box will display with a report showing the results summary
11. Click OK
12. The features that you captured in the field will display

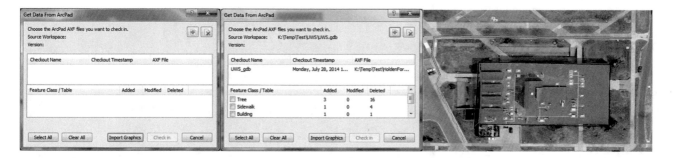

Result: The final map is shown.

Next step:

1. Continue capturing the entire sidewalk around the Holden building
2. Continue capturing the trees around the Holden building
3. Integrate the new capture data
4. Correct the geometry of the building, sidewalk and trees using the image as a reference

Note: When you go to the field and finish capturing the rest of the feature classes (Tree and Sidewalk), you should repeat the steps above and copy the data from ArcPad to your PC folder. Then integrate the data from ArcPad to ArcMap through the "Get Data From ArcPad" on the ArcPad manager. This step will not replace the previous data but will add to it the newly captured data.

Editing the Data of the Holden Building

The data capturing of Holden building using the GPS receiver, shows a displacement due to GPS accuracy. This can be corrected by "start editing" adding some vertex and correct the shape of the building using the image as a background reference.

1. R-click the Building in the TOC/Edit feature
2. D-click on the building feature [4 vertex (green), 1 node (red) and the Edit Vertices dialog box will display]

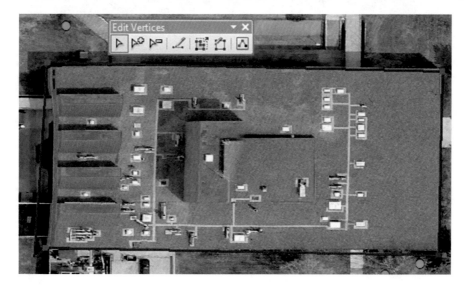

3. Click the Add vertex button on the Edit Vertices to add several vertices
4. Use your GIS skill to move the vertices to the right location to correct the building shape

Note: You might need to add other vertices during editing and make sure to zoom in during the editing process.

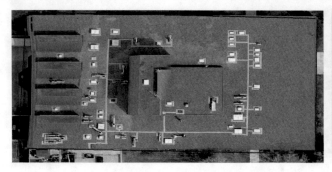 

Quiz: Correct the sidewalk and the location of the trees.

II: QuickProject

A QuickProject is a "ready to use" data capture project. The QuickProject tool creates a "ready to use" data capture project, providing a simple and efficient method for capturing data into three new layers (point, line, and polygon).

Launching ArcPad in Trimble starts with four options:

- New Map
- Create a QuickProject
- Choose map to open
- Open last map used

55. Click Create a QuickProject
56. Template: Default
57. Projection: GCS_WGS_1984
58. Click OK

Result: A new project folder called **QuickProject_YYYYMMDD_HHMM** is created.

Where:

- YYYYMMDD is the current date,
- HHMM is the current time.

If you choose the Default Template and tap OK, the tool creates three new shapefiles in the project folder: Points.shp, Polylines.shp, and Polygons.shp. Each new shapefile has the following fields: Name, Category, Date, Comments, and Photo

Activate the GPS in Trimble

Before using the QuickProject, the GPS with ArcPad should be configured using one of the GPS protocols supported by ArcPad. The GPS Protocol and communication parameters in Trimble Juno T41, is set as follows

ArcPad/GPS Preferences   **SatViewer/GPS/Advanced**

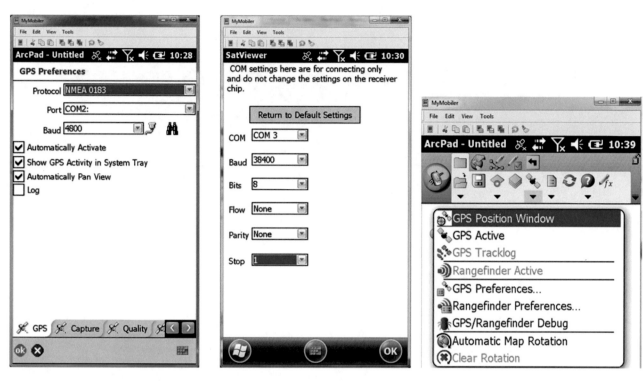

59. Tap the arrow below the GPS Active Icon/tap the GPS Active.

Result: The GPS cursor is displayed when the GPS is active, and is initially located at the current GPS position.

Capturing GPS Data (Point, Line, Polygon)

In this exercise you are going to capture the trees (point), sidewalks (line), and buildings (polygon) at the campus of UWS, as seen in the aerial photograph below. The first step is to capture some of the trees as a point feature. Before you start make sure that your GPS is activated.

Capturing Trees (Point Layer)

60. Find the area where you want to capture the tree
61. Click the Drawing Tools icon
62. Tap the Start/Stop Editing button
63. Tap the Points layer to select it for capturing
64. Tap the point feature tool (3rd icon)

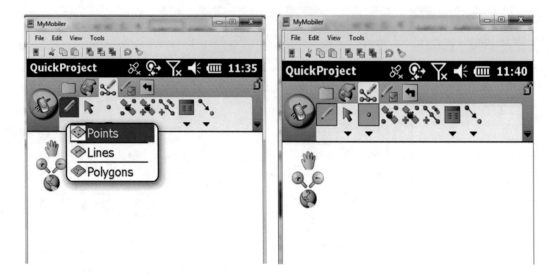

65. Stand close to the tree you are going to capture and click Capture Point (4th icon)
66. The point dialog box will display
67. Name: Tree1
68. Category: select Category 3 (Star)
69. OK

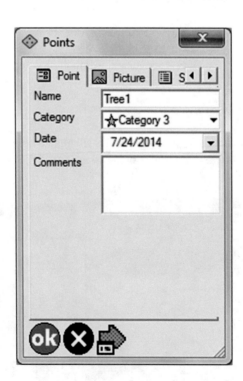

70. Continue to capture the rest of the trees in the designated area
71. OK

Result: The trees are captured in the northern part of Swenson building.

Capturing Sidewalk (Line Layer)
72. Tap the Lines layer to select it for editing.
73. Tap the Polyline feature tool.

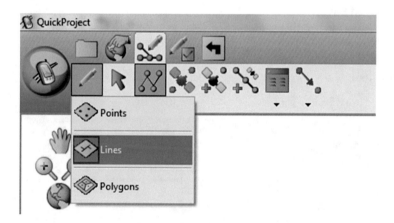

74. Tap the Capture Vertex button (5th icon) at the beginning of Sidewalk to capture a new vertex.
75. Continue to do so until you have reached the end of the Sidewalk.

Note: You can also use the **Capture Vertices Continuously** button (6th icon) to capture vertices in streaming mode. Vertices are captured according to the "specified streaming vertices interval", which is set in the GPS Preferences dialog box.

76. Tap the **Proceed to Attribute Capture tool** (2nd icon) on the Command bar when you are finished adding your Sidewalk.

77. The feature dialog box opens, so that you can immediately enter information about the new Sidewalk.

78. Fill in the form as follows:

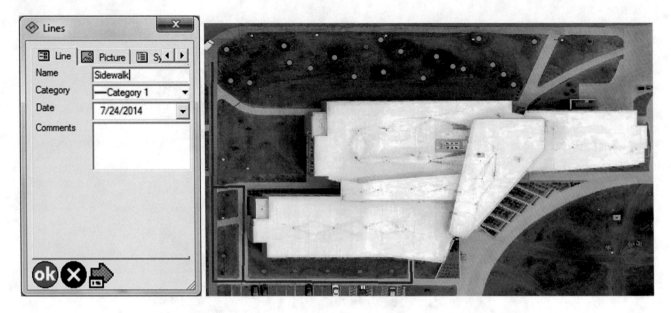

79. OK

Note: Your new Sidewalk has been added, and you are ready to move on to capture the building.

Capturing Building (Polygon Layer)

You are going to capture the Swenson building

80. Tap the Polygons layer to select it for editing.
81. Tap the Polygon feature tool.

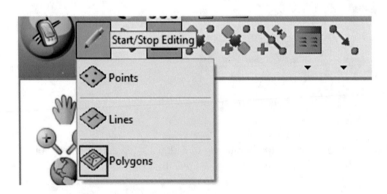

82. Tap the Capture Vertex button (5th icon) at the beginning of the Swenson building to capture a new vertex.
83. Continue to do so until you have reached the end of the building.

Note: You can also use the **Capture Vertices Continuously** button (6th icon), to capture vertices in streaming mode as in the line capturing.

View the Picture That Was Captured in ArcPad in the Field Using Trimble

After using Get Data From ArcPad, fuse the ArPad manager in ArcMap to integrate the **Class_gdb.axf** file.

84. Open the attribute table of the tree

Result: You see under the "Type" field Sample 1, 2, 3. 4, 5, and 6. 4 pictures were taken for Sample 1, 3, 4, and 6.The pictures will be placed under the field "**Photo**".

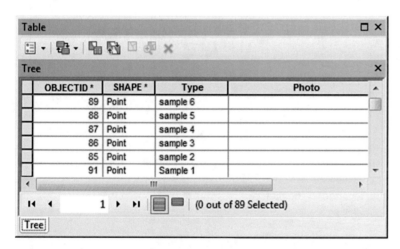

85. In the TOC/R-click Tree/Edit Features/Start Editing
86. Under the **Photo** filed click the row across the sample 1

Result: An Arrow and a line display below the Photo field in the attribute table.

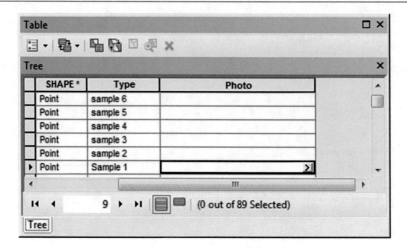

87. Click on the arrow

Result: A window display asking to r-click to load the picture is taken for Sample 1.

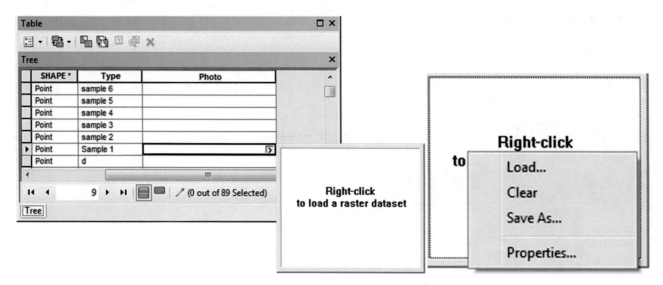

88. R-click inside the window/select load/browse to the location of the picture
89. Fill it as below

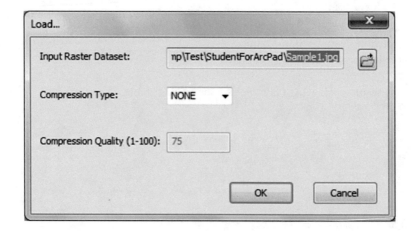

90. OK

Result: The raster is loaded <Raster>, if you click on the arrow the photo will display.

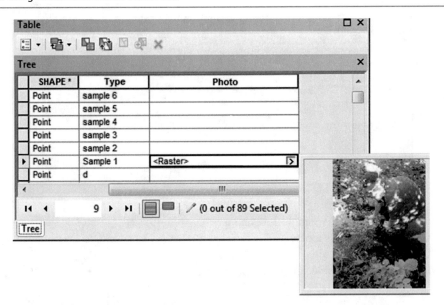

91. Repeat the above procedure for sample 3, 4, and 6
92. Editor Stop Editing

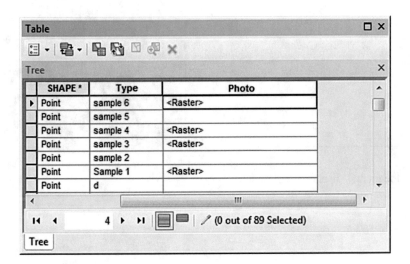

Create Map Hyperlink

93. R-click Tree/label Features
94. Zoom to the Area with Sample 1 to Sample 6
95. Click the Identify tool in the Tools toolbar
96. Click the point marker for Sample 1
97. In the Identify window/R-click Sample 1/Add Hyperlink
98. Click Link to a Document/browse \\StudentForArcPad\ select Sample1
99. Click Open/OK

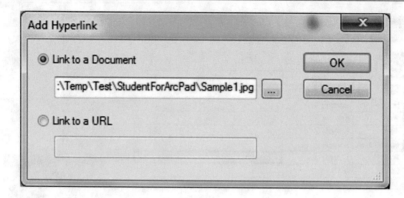

100. Repeat the above procedure to link to Sample 3, 4, and 6
101. Click the Hyperlink tool on the Tools toolbar

Result: All features with hyperlinks available get a blue dot in the center.

102. Click on Sample 1/the picture display

Appendix A: Data Source Credits

Chapter 1: Data source include

Aerial Photograph of Faxon Creek clipped from the "2013 County Aerial 1 Meter Resolution" and downloaded from city of Superior, WI web page: http://www.ci.superior.wi.us/618/Imagery.

Jafr DEM clipped from SRTM 90m Digital Elevation Data downloaded from CGIAR-CSI HOME http://srtm.csi.cgiar.org.

Data (HugIsland, MurphyOil, NewtonCreek, and Sampling layers) are derived from the author article: Assessment of Newton Creek and Its Effect on Hog Island Inlet of Lake Superior. *Journal of Water Quality, Exposure and Health*, 2012 https://link.springer.com/article/10.1007/s12403-012-0071-1.

All graphs are created by the author.

Chapter 2: Data source include

Data (ca_airport, ca_cities, ca_school, and ca_volcanic layers) are taken from ESRI Data and Maps, 2012 https://my.esri.com/#/downloads/Data%20and%20Content.

Data (groundwater wells and pipeline in Texas) created new by the author.

Chapter 3: Data source include

Data (Fault, Geology, and Well layers) are derived from the author article: Recharge origin, overexploitation, and sustainability of water resources in an arid area from Azraq basin, Jordan: case study, 2006 http://hr.iwaponline.com/content/37/3/277.

Chapter 4: Data source include

Data (Douglas, Tract, Lake, and Stream) are derived from TIGER/Line Shapefiles from United States Census Bureau web page: https://www.census.gov/geo/maps-data/data/tiger.html.

GPS data (Coord.txt) created new by students using Garmin GPS designed by the author.

The software (dnrgps.exe) to integrate GPS data into GIS is downloaded from DNR web page: http://www.dnr.state.mn.us/mis/gis/DNRGPS/DNRGPS.html.

Data (World Basemap) is part of ArcGIS Explorer downloaded from ESRI web page: http://www.esri.com/software/arcgis/explorer-desktop.

The image (ortho 1-1 1n s wi031 2015 1.sid) is downloaded from the Wisconsin View web page: ftp://ftp.ssec.wisc.edu/pub/wisconsinview/NAIP_2015/Douglas/.

Data (Protected USA layer) is integrated from ArcGIS Online https://www.arcgis.com/home/index.html.

Data (Catch and Well layers) are derived from the author article: Recharge mechanism and hydrochemistry evaluation of groundwater in the Nuaimeh area, Jordan, using environmental isotope techniques, 2006 http://link.springer.com/article/10.1007/s10040-004-0352-2.

Data (points, lines, polygons) are created new by students using on screen digitizing designed by the author.

The image (UWS_UTM15.tif) is clipped and projected from the "2013 County Aerial 1 Meter Resolution" and downloaded from city of Superior, WI web page: http://www.ci.superior.wi.us/618/Imagery.

© Springer International Publishing AG 2018

W. Bajjali, *ArcGIS for Environmental and Water Issues*, Springer Textbooks in Earth Sciences, Geography and Environment, https://doi.org/10.1007/978-3-319-61158-7

Chapter 5: Data source include

Data (streets of Superior, street of Duluth, and WI layers) are derived from TIGER/Line Shapefiles from United States Census Bureau web page: https://www.census.gov/geo/maps-data/data/tiger.html.

Data (Lake.tif clipped from ortho_1-1_1n_s_wi078_2015_1.sid) downloaded from the Wisconsin View web page: ftp://ftp.ssec.wisc.edu/pub/wisconsinview/NAIP_2015/Menominee.

Data (GPS layer) created new by the author using Garmin GPS.

The image (Aquifer.jpg) represents the Eastern dolomite aquifer of Wisconsin is downloaded from the web page: http://wgnhs.uwex.edu/water-environment/wisconsin-aquifers/.

Data (Well table and Dhuleil layer) are derived from the author article: Water quality and geochemistry evaluation of groundwater upstream and downstream of the Khirbet Al-Samra wastewater treatment plant/Jordan 2015 http://link.springer.com/article/10.1007/s13201-014-0263-x.

Data sketch No. 2, 3, and 7 are taken from ESRI web page: http://desktop.arcgis.com.

Chapter 6: Data source include

The image (Rec_Stream.tif) is derived from the author article: Recharge origin, overexploitation, and sustainability of water resources in an arid area from Azraq basin, Jordan: case study, 2006 http://hr.iwaponline.com/content/37/3/277.

Chapter 7: Data source include

Data (image_rectify, Dam, Stream, and StudyArea layers) are derived from the author article: Water quality and geochemistry evaluation of groundwater upstream and downstream of the Khirbet Al-Samra wastewater treatment plant/Jordan, 2015 http://link.springer.com/article/10.1007/s13201-014-0263-x.

Data (Well, Fault, and Plant layers) are created new using geodatabase forms designed by the author.

Data (Well, Catchment, Table 1, and Table 2 layers) are created new by the author.

Data (Well and Dhuleil layers) are derived from the author article: Water quality and geochemistry evaluation of groundwater upstream and downstream of the Khirbet Al-Samra wastewater treatment plant/Jordan 2015 http://link.springer.com/article/10.1007/s13201-014-0263-x.

Chapter 8: Data source include

Data (Farm and LandB layers) created new by the author for editing purposes.

Data (Street_Mn, River_MN, Lake_MN, Fault, Watershed_1, and Watershed_2 layers) created new by the author for editing purposes.

Data (Geology and Field_Geology layers) are derived from the author article: Water quality and geochemistry evaluation of groundwater upstream and downstream of the Khirbet Al-Samra wastewater treatment plant/Jordan, 2015 http://link.springer.com/article/10.1007/s13201-014-0263-x.

The image (North_Duluth.jpg) is an aerial photograph of St. Louis County derived and downloaded from MN-DNR web page: http://www.dnr.state.mn.us/maps/landview.html.

Chapter 9: Data source include

Data (watershed, stream, soil, FarmA, FarmB, FarmC, and FarmD layers) are created new by the author.

Chapter 10: Data source include

Data (Landuse, Pipeline, StudyArea, and Vegetation layers) are digitized on screen from an image representing University of Jordan. The image had been downloaded from Google Earth (https://www.google.com/earth). The image then clipped and georeferenced using local projection.

Data (Well, Fault, Stream, KSWTP, and GEOL_KS layers) are derived from the author article: Water quality and geochemistry evaluation of groundwater upstream and downstream of the Khirbet Al-Samra wastewater treatment plant/Jordan, 2015. http://link.springer.com/article/10.1007/s13201-014-0263-x.

Chapter 11: Data source include

Data (ZipCode_WI and Street of city of superior layer) are derived from ESRI Data and Maps, 2012 https://my.esri.com/#/downloads/Data%20and%20Content.

Data (Well_Owner layer and Well table) are created new by the author for the purpose of Geocoding.

Chapter 12: Data source include

The elevation image of city of Superior in Wisconsin (USGS_NED_13_n47w093_ArcGrid.zip) is downloaded from the USGS web page: http://viewer.nationalmap.gov/.

The images (AZ_DEM, Dhuleil.tif and KTDam grids) are downloaded and clipped from SRTM 90m Digital Elevation Data downloaded from CGIAR-CSI HOME http://srtm.csi.cgiar.org.

Data (Geology, stream, and Luhfi_Dam layers) are derived from the author article: Water quality and geochemistry evaluation of groundwater upstream and downstream of the Khirbet Al-Samra wastewater treatment plant/Jordan, 2015. http://link.springer.com/article/10.1007/s13201-014-0263-x.

Chapter 13: Data source include

Data (Dam, Stream, Watershed and Well layers) are derived from the author article: Model the effect of four artificial recharge dams on the quality of groundwater using geostatistical methods in GIS environment, Oman, 2005 http://www.spatialhydrology.net/index.php/JOSH/article/view/39.

Chapter 14: Data source include

Data (DEM and layer of Dhuleil) are derived from the author article: Water quality and geochemistry evaluation of groundwater upstream and downstream of the Khirbet Al-Samra wastewater treatment plant/Jordan, 2015. http://link.springer.com/article/10.1007/s13201-014-0263-x.

Chapter 15: Data source include

Data (Governorate, Town, Well, WalaWatershed, Grid_1000, Geology, and WWTP layers) are derived from GIS Workshop in Hydrogeology at Water Authority of Jordan, Ministry of Water and Irrigation, September 2-6, 2012 created and instructed by the author.

Data (Dam, Geology, Stream, and Well layers) are derived from the author article: Water quality and geochemistry evaluation of groundwater upstream and downstream of the Khirbet Al-Samra wastewater treatment plant/Jordan, 2015. http://link.springer.com/article/10.1007/s13201-014-0263-x.

Chapter 16: Data source include

Data (Dam, Region, Road, Stream, Street, Town, Well, Well_Supply, and WWTP layers) are derived from GIS Workshop in Hydrogeology at Water Authority of Jordan, Ministry of Water and Irrigation, September 2–6, 2012 created and instructed by the author.

Chapter 17: Data source include

The DEM of city of Duluth, MN (Duluth.tif) is downloaded from the USGS web page: http://viewer.nationalmap.gov/. The data (Contour and Duluth layers) are obtained from the DEM (Duluth.tif).

Data (City and Stream layers) are taken from ESRI Data and Maps, 2012 https://my.esri.com/#/downloads/Data%20 and%20Content.

Data (RainStation layer) is created new by the author.

Data (Building, Farm, GasStation, ObserbationWell, Plume, Street, SupplyWell, Tree, Valley, and WWTP layers) are derived from GIS Workshop in Hydrogeology at Water Authority of Jordan, Ministry of Water and Irrigation, September 2–6, 2012 created and instructed by the author.

Chapter 18: Data source include

The aerial photograph images (UWS_GCS.tif and Holden.tif) are clipped from the image "2013 County Aerial 1 Meter Resolution" downloaded from the city of Superior web page: http://www.ci.superior.wi.us/618/Imagery.

Data (Building, Sidewalk, and Tree feature classes) are created new by students using on screen digitizing designed by the author.

Appendix B: Task-Index

No	Task-Index	Chapter
1	1-Band	1, 4, 6, 12, 17
2	3-D Analysis	17
3	3-D Symbol	17
4	3D Thumbnail	17
5	3-D Toolbar	17
6	3D Trees	17
7	Add Data	1, 2, 4, 6, 12
8	Add Field	4, 9, 10, 12, 14, 15, 16
9	Add Layer	18
10	Add Locations Tool	16
11	Add Rule	8
12	Address Locator	11
13	Address Table	11
14	Advance Editing	8
15	Aerial Photography	4
16	Analysis Tool	9, 10, 16
17	Analyzing Patterns	15
18	Animation Control	17
19	Animation toolbar	17
20	Anselin Local Moran I	15
21	Append	9
22	ArcCatalog	2, 5, 7, 8, 17
23	ArcGrid	12
24	ArcMap	1–17
25	ArcPad	18
26	ArcPad Data Manager tool	18
27	ArcScan	6
28	ArcScene	17
29	ArcToolbox	2, 5, 6, 18
30	Aspect	12
31	Assign Coordinate	5
32	Attribute Table	1–18
33	Auto Complete Polygon	4
34	Auto Layout	10
35	AutoCAD	7
36	Automatic Vectorization	6
37	Average Nearest Neighbor	15
38	Base Height	17
39	Bi-level image	8
40	Bookmark	2, 8

(continued)

© Springer International Publishing AG 2018
W. Bajjali, *ArcGIS for Environmental and Water Issues*, Springer Textbooks in Earth Sciences, Geography and Environment,
https://doi.org/10.1007/978-3-319-61158-7

No	Task-Index	Chapter
41	Buffer	9, 10, 15, 16
42	Calculate Distance Band from Neighbor Count	15
43	Calculate Geometry	8, 9, 10, 14
44	Capture GPS Data	18
45	Catalog window	4, 5, 6, 7, 8, 10, 11, 16, 18
46	Census Tract	4
47	Central feature	15
48	Change Name	2
49	Classification	2, 6, 9, 12, 16
50	Classify	2, 3, 5, 6, 8, 9, 11, 12, 13, 16
51	Clear Selected Feature	18
52	Clip	9, 12, 13
53	Closest Facility	16
54	Cluster and Outlier Analysis	15
55	Combine	9
56	Combining features	9
57	Combining geometries and attributes	10
58	Con tool	14
59	Confidence Level	15
60	Connect	1–8, 10, 17, 18
61	Construction Tools	4, 6, 7, 8
62	Contour	6, 12, 17
63	Control Points	5, 13
64	Conversion tool	11
65	Convert	4, 9, 12, 13, 14
66	Convert Graphic to Feature	9
67	Coordinate Systems	5, 7, 8, 12, 14, 18
68	Copy	1–5, 10, 12, 13, 17, 18
69	Create Animation	17
70	Create Feature Dataset	7
71	Create Geodatabase	7, 8
72	Create Layer	17
73	Create Line Shapefile	4, 6, 7
74	Create Point Shapefile	4, 7
75	Create Polygon Shapefile	4, 7, 18
76	Create Relationship Class	7
77	Create TIN	12, 17
78	Create Video	17
79	Customize menu	3, 5, 6
80	Cut Polygon Tool	8
81	Data Acquisition	4
82	Data Editing	8
83	Data Integration	2–4, 6, 18
84	Data Management Tools	5, 9
85	Data View	2, 4, 5, 12
86	Database Management Systems (DBMS)	9
87	Datum	1, 5, 7, 8
88	Datum Conflict	4, 5, 7, 12
89	Decimal Degree	5
90	Define Projection	5
91	Definition Query	15, 16
92	Delete	2–6, 8, 10–12
93	DEM	1, 12, 14, 17

(continued)

No	Task-Index	Chapter
94	Density	1, 5, 13, 15
95	Desire Lines (Spider Diagram)	16
96	Digitize	1, 4, 6, 7, 8, 10, 17, 18
97	Digitizing on Screen	4, 8, 10, 17, 18
98	Dissolve	4, 5, 10, 11, 12, 18
99	Download GIS Data	12
100	Drape	17
101	Draw tool	3, 9
102	Edit Feature	4, 5, 6, 8, 18
103	Editor Toolbar	4–8
104	Elevation Products (3DEP)	12
105	Equal Intervals Method	3
106	Erase	6, 9, 10
107	Erase tool	6, 9, 10
108	Euclidean Distance	15, 16
109	Excel	1, 4, 5, 11
110	Export Animation	17
111	Export Map	18
112	Export to Raster	13
113	Extend tool	8
114	Extract by Attribute	12
115	Extracting Features	9
116	Extrude	17
117	Faces with the same symbol	17
118	Feature Class	1, 2, 4, 5, 7, 8, 9, 10, 15, 16, 18
119	Feature Creating	4
120	Feature Dataset	7, 8, 16, 18
121	Field Calculator	9, 12, 16
122	File Geodatabase	7, 8, 18
123	Find	2, 4, 7, 9, 10, 12, 15–18
124	Fix Lines using topology	8
125	Fix Polygons using topology	8
126	Fix Topology Error Tool	8
127	Fixing Dangle	8
128	Floating on a Custom Surface	17
129	Flow Accumulation	14
130	Flow Direction	14
131	Fly	17
132	Full Extent	2, 3, 5, 8, 9, 10, 12, 17, 18
133	Garmin Extension	4
134	Generalize feature	8
135	Generate Features	6
136	Geocoding	11
137	Geodatabase	1, 2, 7, 8, 18
138	Geodatabase File	7, 8, 18
139	Geodatabase Personal	7, 16
140	Geographic Coordinate System (GCS)	4, 5, 7, 12, 14, 18
141	Geoprocessing	9
142	Geoprocessing Environment	9
143	Geoprocessing Options	10
144	Georeferencing	5
145	Geostaistical Wizard	13
146	Geostatistical Analyst	13

(continued)

No	Task-Index	Chapter
147	GEOTif	18
148	Get Data For ArcPad	18
149	Getis-Ord General G	15
150	Getis-Ord GI*	15
151	Global Polynomial (GP)	13
152	Global Positioning System (GPS)	1, 4, 5, 8, 18
153	GPS Preferences	18
154	Graduate symbols	11, 13, 16
155	Greenhouse	4, 10
156	Guidelines	3
157	Hard Line	17
158	Hillshade	12
159	Hot Spot Analysis	15
160	Hydrology	1, 8, 14
161	Identify	2, 14, 18
162	Import	4–9, 12, 16–18
163	Index New Items	2
164	Insert	1, 3, 4, 5, 7, 8, 9, 11–17
165	Int tool	12
166	Interactive Vectorization	6
167	Intermediate Data	10
168	Interpolate Line	12, 17
169	Interpolate Point	17
170	Interpolation	13
171	Intersect	9, 16
172	Inverse Distance Weighting (IDW)	13, 14
173	Join	4, 7, 8, 10, 15, 16
174	Join and Relate	4, 7, 8, 10, 15, 16
175	Kriging	13, 14
176	Label	2–6, 8–10, 12–13, 16–18
177	Landscape	1, 3, 4, 12
178	Latitude-Longitude	1, 4, 5, 7, 12, 18
179	Layer Package (LPK)	4
180	Layout View	2, 3
181	Legend	3, 4, 6
182	Line of Sight	12, 17
183	List by Drawing Order	4
184	Magic Erase tool	6
185	Main Tools	18
186	Manual Method	3
187	Map Document	2, 3, 16
188	Map Hyperlink	18
189	Map Projections	5, 9, 12–14
190	Mapping Clusters	15
191	Mask	13
192	Match	11
193	Match_Addr	11
194	Mean	9, 11–13
195	Mean Center	15
196	Measuring Geographic Distributions	15
197	Median Center	15

(continued)

No	Task-Index	Chapter
198	Mercator	5, 6, 7, 12
199	Merge	8, 9, 12
200	Meridians and Parallels	5
201	Metadata	2, 5
202	Mobile GIS	18
203	Model	10
204	Model Parameters	10
205	ModelBuilder	9, 10
206	Modify	4, 8, 10
207	Moran's I	15
208	Mosaic	12
209	Move	8
210	Multi-Ring Buffer	16
211	Must Not Have Dangle	8
212	Nationalmap.gov	12
213	Natural Breaks (Jenks)	3
214	Navigate	17
215	Near	15, 16
216	Nearest Neighbor Index (I)	15
217	NED	12
218	Neighborhood	13–16
219	Network Analyst	16
220	Network Dataset	16
221	North American Datum (NAD)	5, 8
222	Null Hypothesis	15
223	Ortho	4
224	Overlay Analysis	9, 10
225	Overshoots	8
226	Page and Print Setup	3
227	Percent_Rise	12
228	Point Distance	16
229	Polygon to Raster	12
230	Primary Table	11
231	Project Raster	5, 12
232	Projection	1, 2, 5, 7, 10, 12, 17–18
233	Projection (Azimuthal or True-Direction)	5
234	Projection (Conformal or Orthomorphic)	5
235	Projection (Equidistant)	5
236	Projection (Equivalent or Equal-Area)	5
237	Projection and Distortion	5
238	Projection on the Fly	5
239	Projection Parameters	5
240	Projections and Transformations	5, 12, 14
241	Proximity	5, 9, 10, 13, 15–16
242	Quantile Method	3
243	Query Builder	15, 16
244	Quick Capture toolbar	18
245	Quick Project	18
246	Raster	1, 5, 6, 12, 14, 16–18
247	Raster Calculator	12

(continued)

No	Task-Index	Chapter
248	Raster Cleanup	6
249	Raster Conversion	12
250	Raster Dataset	1, 5, 7, 12, 18
251	Raster Information	1, 12
252	Raster Painting toolbar	6
253	Raster Projection	5, 6, 12, 18
254	Ready-to-run	10
255	Reclassify tool	6, 12
256	Reference Data	11
257	Relational Database	7
258	Relationship class	7
259	Relative Paths	2
260	Rematch	11
261	Rename	2, 3
262	Resample	12
263	Reshape	8
264	Root Mean Square (RMS)	5
265	Run model	10
266	Satellite Images	4, 5
267	Scale bar	3
268	Scale Map Element	3
269	Search window	10
270	Select by attributes	7, 8
271	Select by location	9, 16
272	Select tool	10
273	Semimajor and Semiminor Axes	5
274	Service Area	16
275	Set the Environment	5, 6, 9, 10, 12–16
276	Shapefile	1, 3–9, 11–13, 17–18
277	Show Map Tips	2
278	Simple Editing	8
279	Sink	14
280	Site Suitability	10
281	Slope	12
282	Smooth feature	8
283	Snapping Toolbar	5, 6, 8
284	Sort	3, 11
285	Spatial Analysts	6, 8, 12, 13, 14, 16
286	Spatial Autocorrelation (Moran's I)	15
287	Spatial Join	9, 15, 16
288	Spatial Reference	12
289	Spatial Statistics	15
290	Spherical Coordinate System	5
291	Spider Diagram (Desire Lines)	16
292	Split	8, 9
293	SQL	7, 8, 10, 12, 14–16
294	Standard Distance	15
295	State Plane Coordinate	5
296	Statistics	1, 9, 12, 13, 15–17
297	Stop Editing	4, 6, 8, 18

(continued)

No	Task-Index	Chapter
298	Straight Segment	8
299	Stream Feature	6–9
300	Stream Link	14
301	Style References	3, 17
302	Subtract	9
303	Sum	9
304	Summarize	11, 15
305	Switch	16, 18
306	Symbolize	1–3, 13, 14, 16
307	Table Options	4, 7, 8, 12, 16
308	Terrain Analysis	12
309	Thumbnail	2, 17
310	Tiger Products	4
311	Time Slider	17
312	Time Tracking	17
313	Tools toolbar	2, 4, 5, 9, 12, 17, 18
314	Topological Editing	8
315	Topology	8
316	Transparency	12, 13, 17
317	Trend Surface Analysis	13
318	Triangulated Irregular Network (TIN)	17
319	Trim tool	8
320	Trimble Juno	18
321	Undershoots	8
322	Union	9, 10
323	US Address—Dual Ranges	11
324	Use Snapping	6, 8
325	USGS	11, 12
326	UTM	1, 4–5, 7–9, 12–13
327	Validate Entire Model	10
328	Validate Topology on Current Extent	8
329	Vectorization	6
330	Vectorization Setting	6
331	Vectorization Trace	6
332	Vertical Exaggeration	17
333	Vertical Profile	12, 17
334	Visibility map	12
335	Watershed	4, 7–9, 13–15
336	Waypoints	4
337	Weight	15
338	Workspace	5, 6, 8–16
339	World geodetic System (WGS)	5, 7–9, 12–14, 18
340	Write Geo TIFF Tags	18
341	Write World File	18
342	Zip Code	11
343	Z-score	15